Teleportation: From Star Trek® To Tesla

COMMANDER X

With Tim Swartz

☆ ☆ ☆

Global Communications

Order all your Tesla products from:
www.TeslaSecretLab.com

24 hour automated order hot line (732) 602-340

TELEPORTATION: A HOW-TO GUIDE -- FROM STAR TREK® TO TESLA
All Rights Reserved - copyright© 2001
Global Communications/Timothy G Beckley

●●●●●●●●●●●●●●●●●●●●●●●●●●●●●●●●●●●●

ISBN: 1-892062-43-7
Science/New Age/Alternative Technology

Written by Commander X with Tim Swartz

Timothy G. Beckley, Editorial Director
Carol Rodriguez, Publisher's Assistant

Cover art and graphics by Chris Flemming

Free catalog from Global Communications,
Box 753, New Brunswick, NJ 08903
Free subscription to CONSPIRACY JOURNAL
e-mailed every week. Go to below site NOW.
www.conspiracyjournal.com

Contents

Introduction By Commander X   **5**

Chapter One: Does Teleportation Exist?   **10**

Chapter Two: Science Fiction–Science Fact   **20**

Chapter Three: Beam Me Up Scotty!   **29**

Chapter Four: Gifts From The World of Spirits   **33**

Chapter Five: Pennies From Heaven   **42**

Chapter Six: Recollections Of Area 51   **49**

Chapter Seven: Secret Research In Spacetime Travel   **60**

Chapter Eight: UFO Abductions – The Teleportation Factor   **69**

Chapter Nine: The Weird World Of Teleportation   **81**

Chapter Ten: Uri Geller And The Superminds   **93**

Chapter Eleven: The Tesla Teleportation Project   **102**

Chapter Twelve: Practical Teleportation   **111**

Chapter Thirteen: Personal Experiences In Teleportation   **128**

Chapter Fourteen: The Teleportation Of Sacred Stones
By Diane Tessman   **132**

ACKNOWLEDGMENTS

The authors wish to acknowledge the following individuals without whose help this book would not have been possible: Nicky Malloy - Diane Tessman - Peter Robbins - Tim Beckley - Carol Rodriguez - Lee Hefley - Ron Bonds - and of course, Mom, Dad and Melissa.

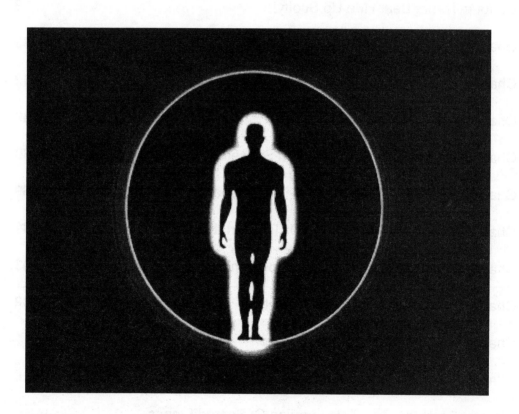

Teleportation: From Star Trek To Tesla

INTRODUCTION

"Beam me up, Scotty!" This phrase from the 1960's television show *Star Trek* has become synonymous in our era of pop entertainment for the concept of teleportation. **Teleportation**, the instantaneous movement of an object from one location to another without passing through the space between.

The idea of teleportation was originally thought of in the 1930's when Albert Einstein, Boris Podolsky, and Nathan Rosen wrote a famous paper concerning the scanning process that would be involved during teleportation, this was later known as the EPR effect. However, the actual term, teleportation, was originally conceived by none other than writer Charles Fort in his fascinating book *LO!*, published back in 1931.

The word appears at the end of chapter two following a discussion of a great fall of periwinkles in Worcester U.K. in 1881. Fort discounts wind and fishmongers throwing away winkles as a source of the creatures, and he writes a short philosophical chapter proposing a somewhat tongue in cheek Gaia-like idea with the world as a sort of organism having a mechanism for transporting organisms and materials as falls to where they are required and he concludes: "Mostly in this book I shall speculate upon indications that there exists a transportory force that I shall call teleportation."

This is apparently the first use of the word and it soon became known to science via science fiction. Ray Palmer, editor of *Astounding* magazine in the 1940's, used to encourage writers to read Fort for plot ideas, and some picked up on the term teleportation, and it has remained with us ever since.

Even though the concept for teleportation was popular in science fiction, most scientists considered the reality of such a feat to be totally impossible. After all, to be able to build a transporter like the ones popularized in *Star Trek* and countless other books, movies and TV shows, would require the storage capacity of over 100,000,000,000,000,000,000,000,000 (100 septillion) atoms in the human body.

The next major step would be to scan all the atoms in a human body. To check their position, charge, orientation, quantum states, etc. As far as we know, this can't be done, yet. One problem is that scanning would change the quantum state of a person's atoms to such an extent that reassembly would be impossible.

But if we could scan all the atoms and their properties in a body without changing them with the scan, we would then have to figure out a way to harmlessly disintegrate a person, then transport all their atoms to another place and then perfectly reassemble them again without any glitches or the computer crashing. Certainly an unpleasant prospect considering how often modern programs crash and burn during a normal day's operations.

Despite the daunting task, as well as the perceived taint of science fiction, work on forms

Teleportation: From Star Trek To Tesla

of teleportation is currently underway in labs all across the globe. In 1997, a team of scientists at the University of Innsbruck destroyed bits of light in one place and made perfect replicas appear about three feet away. They accomplished this by transferring information about a crucial physical characteristic of the original light bits, called photons.

The information was picked up by other photons, which took on that characteristic and so became replicas of the originals. The phenomenon that made it happen is so bizarre that even Albert Einstein didn't believe in it. He called it spooky. Besides raising the rather fantastic notion of a new means of transportation the trick could lead to ultra-fast computers.

The experiment was reported in the journal *Nature* by Anton Zeilinger and colleagues at the University of Innsbruck in Austria. Another research team, based in Rome, has done similar work and submitted its report to another journal.

The work is the first to demonstrate "quantum teleportation," a bizarre shifting of physical characteristics between nature's tiniest particles, no matter how far apart they are. Scientists might be able to achieve teleportation between atoms within a few years and molecules within a decade or so, Zeilinger said.

Until recently, teleportation was not taken seriously by scientists, because it was thought to violate the uncertainty principle of quantum mechanics, which forbids any measuring or scanning process from extracting all the information in an atom or other object. According to the uncertainty principle, the more accurately an object is scanned, the more it is disturbed by the scanning process, until one reaches a point where the object's original state has been completely disrupted, still without having extracted enough information to make a perfect replica.

This sounds like a solid argument against teleportation; if one cannot extract enough information from an object to make a perfect copy, it would seem that a perfect copy cannot be made. But scientists have found a way to get around this logic by using a celebrated and paradoxical feature of quantum mechanics known as the Einstein-Podolsky-Rosen effect.

In brief, they found a way to scan out part of the information from an object A, which one wishes to teleport, while causing the remaining, unscanned, part of the information to pass, via the Einstein-Podolsky-Rosen effect, into another object C which has never been in contact with A.

Later, by applying to C a treatment depending on the scanned-out information, it is possible to maneuver C into exactly the same state as A was in before it was scanned. A itself is no longer in that state, having been thoroughly disrupted by the scanning, so what has been achieved is teleportation, not replication. The unscanned part of the information is conveyed from A to C by an intermediary object B, which interacts first with C and then with A.

There is a subtle, unscannable kind of information that, unlike any material cargo, and even unlike ordinary information, can be delivered in a backward fashion. This subtle kind

Teleportation: From Star Trek To Tesla

of information, also called "Einstein-Podolsky-Rosen (EPR) correlation" or "entanglement," has been at least partly understood since the 1930s when it was discussed in a famous paper by Albert Einstein, Boris Podolsky, and Nathan Rosen.

In the 1960s, physicist John Bell showed that a pair of entangled particles, which were once in contact but later move too far apart to interact directly, can exhibit individually random behavior that is too strongly correlated to be explained by classical statistics. Experiments on photons and other particles have repeatedly confirmed these correlations, thereby providing strong evidence for the validity of quantum mechanics, which neatly explains them.

Another well-known fact about EPR correlations is that they cannot by themselves deliver a meaningful and controllable message. It was thought that their only usefulness was in proving the validity of quantum mechanics. But now it is known that, through the phenomenon of quantum teleportation, they can deliver exactly that part of the information in an object which is too delicate to be scanned out and delivered by conventional methods.

The underlying principle is fundamentally different from the Star Trek process of beaming people around. But could this kind of teleportation be used on people? Could scientists extract information from every tiny particle in a person, transfer it to a bunch of particles elsewhere, and assemble those particles into an exact replica of the person?

There's no theoretical problem with that, several experts have stated. "I think it's quite clear that anything approximating teleportation of complex living beings, even bacteria, is so far away technologically that it's not really worth thinking about it," said IBM physicist Charles H. Bennett. He and other physicists proposed quantum teleportation in 1993.

There would just be too much information to assemble and transmit, several researchers have noted. Even if it were possible someday, it would be so expensive that "probably it's just as cheap to send the real person," said Benjamin Schumacher of Kenyon College in Gambier, Ohio. Besides, Schumacher said, teleportation would "kill you and take you apart atom by atom so you could be reassembled at the other end, one hopes. It doesn't seem like a good idea to me."

Much more likely, experts maintain, is using teleportation between tiny particles to set up quantum computers. These devices would use teleportation to sling data around, and they could solve certain complex problems much faster than today's machines.

Marc Millis, Aerospace Engineer, at the NASA Lewis Research Center, is currently the leader of a group of about 150 scientists and engineers that represents a program at NASA known as The Breakthrough Propulsion Physics Program. Marc had this to say regarding teleportation: "Transporters are only at the very first step of evolution, – conjecture. We know what we'd like to accomplish, but we haven't got a clue what physics we'll even need to learn about. The closest topic to a transporter that we do have some understanding of is wormholes. Because this subject is much less developed than the search for gravity control or hyperfast propulsion, there is little or no work being directly aimed at the question of transporters."

I know for a fact that teleportation is a reality, that it is possible to transport oneself in the twinkling of an eye. I know because I witnessed one of the various forms of teleportation first hand while in the military, and I know for an absolute fact that our government – or secret branches of it I should say – has perfected an aircraft that can literally be here one moment and somewhere else an eye blink later.

At Area 51 I have watched from inside the perimeter of Groom Lake, while dozens of skywatchers and tourists crane their necks in the dark looking toward the heavens, as they stand near what has become known as the black mailbox, the best viewing point available on private property.

In the silence of the night huge bunker doors would open up from the hillside and a craft would emerge from underground and lift slowly, and silently, into the blackness. Once aloft at 20 or 30 thousand feet and at what should be a "safe distance" from prying eyes, a computer command onboard this sleek metallic ship would shift the craft into warp speed where it would actually be able to move from "here to there" in literally nothing flat.

For while no one has released this information on the six o clock news, and it would be denied by the Pentagon that any such mod of transport exists outside the realm of science fiction, teleportation is a technology that is very much being tested today and has been successful.

While I was never privy to any real information concerning the true nature of these amazing craft – I was told stories about Grey aliens and Nordics battling for our planet, and how our military had captured alien spacecraft. Stories that could never be verified and were possibly disinformation to keep the likes of me quiet. Nevertheless, physicist Bob Lazar claims that he does know how these craft operate, and it involves the warping of time and space.

Lazar says that the disc's reactor uses a fuel which does not occur naturally on Earth. This fuel is a super-heavy, stable, element with an atomic number of 115 and does not appear on our periodic chart. Element 115 has a twofold purpose : First, it is the source of a gravity wave that is unknown to Earth's scientists, the "Gravity A" wave.

Second, it is the source of the anti-matter radiation which is reacted to provide power. The Gravity A wave emanates from the nucleus of Element 115 and actually extends past the perimeter of the atom. The propulsion system of the disc amplifies and focuses this Gravity A wave to cause space/time to bend, much like space/time bends in the intense gravitational field of a black hole.

The ability to direct gravity to cause space/time distortions allows the disc to cross vast expanses of space/time without traveling in a linear mode at a high rate of speed. Inside the reactor, the Element 115 is transmuted to Element 116 which is unstable and immediately

Teleportation: From Star Trek To Tesla

decays releasing antimatter. The antimatter is reacted with gaseous matter which causes a total annihilation reaction, the 100 percent conversion of matter to energy. The heat from this reaction is converted to electrical energy by a solid state, near 100 percent efficient thermoelectric generator. It is this energy that is used to amplify the Gravity A wave.

It allows you to do virtually anything. Gravity distorts time and space. By doing that, now you're into a different mode of travel, where instead of traveling in a linear method – going from Point A to B – now you can distort time and space to where you essentially bring the mountain to Mohammed; you almost bring your destination to you without moving.

Since you're distorting time, all this takes place in between moments of time. On the bottom side of the disc are the three gravity generators. When travel is desired to a distant point, the disc turns on its side. The three gravity generators will produce a gravitational beam.

What then happens is the gravity generators are converged onto a point which is used as a focal point. The generators are brought up to power and the point is pulled towards the disc. The disc itself will attach onto that point and snap back – as space is released back to its original location. If you are on the ground watching such a performance, the craft would appear to disappear and reappear almost instantaneously in a different part of the sky. In essence, a form of teleportation.

Scientists using what little they understand of this incredible technology have tried to replicate it in a laboratory without the spacecraft. Their ultimate goal is to create a method of teleportation just like you see on Star Trek, where you can "beam" yourself to some other location. So far, the tests have not been successful, and in fact, have led to some devastating failures with an appalling loss of some brilliant people.

So it seems that science still has a long way to go before we can safely send our atoms streaming across the universe to holiday on some alien beach in another galaxy. Nevertheless, stories abound in of objects, animals, and even people being mysteriously transported from one location to another. So there could be some natural process allowing for teleportation that our current science has either overlooked, or has yet no concept. A process that may be easily tapped into simply by having the right state of mind, and being in the right place at the right time.

It may be centuries before we can say: "Beam me up Scotty!" and find ourselves flying into the cosmos like radio waves. However, we may not have to wait that long if we make a serious effort to study and understand phenomena such as apports, strange disappearances and appearances, and even UFO abduction encounters. A whole new understanding of our reality may be tantalizingly close. Yet, still just out of our reach.

Commander X

CHAPTER 1
DOES TELEPORTATION EXIST?

For Laurie Nelson it was a night like any other night as she drove home from her job at a local convenience store in Ames, Iowa. The road to her rural home was dark and curvy, but she had taken this route hundreds of times before and knew it well enough to drive even under the most extreme of conditions.

The night of August 16, 1985 offered nothing out of the ordinary for the tired woman, who was anxious to finally get home and end her day with a hot cup of tea. However, this night would forever change how Laurie Nelson viewed her world, and how a quiet, unassuming evening could suddenly change with terrifying results.

As Laurie drove down the dark, country road, she realized that it had been awhile since she had passed any other car. This was odd she thought as she often encountered other people from the area heading for their night-shift jobs at area factories. However, tonight was different. No headlights from oncoming cars broke the tedium of Laurie's drive home.

Laurie also noticed that the night seemed darker than normal. There was no moon out as the sky was covered by high clouds. But the air seemed denser, much like a dark blanket that surrounded her car. She also noticed that her headlights seemed dimmer than normal. It almost seemed as if the lights were cutting through empty, dark space that was loathe to give up its hold to the glaring lights from her car. She looked at her watch – it was twenty after ten.

As the car sped around a dog leg, it was suddenly confronted by what seemed to be a fog bank lying directly over the road. Before she could even slow down, Laurie was completely surrounded by the fog which seemed to be entering the car through the rolled down windows. As the strange fog filled the inside of the car, Laurie noticed that the speedometer was registering in excess of sixty miles an hour, but to the frightened women, it felt as if the car had come to a dead stop. Stranger still, she realized she couldn't hear the engine or any other sound for that matter. It was as if she had suddenly lost all ability to hear.

The next thing the terrified woman realized was that she was now standing in a flat field of low grass. It was pitch black, her car and the strange fog bank were nowhere in sight. She was alone and utterly frightened.

Her first thought was that she had been in a wreak and had been thrown clear of her car. But the night was dark and quiet and she appeared uninjured. She pressed the switch of her watch and the face illuminated showing her that the time was now 10:22. Whatever had happened to her had occurred within the last two minutes.

Slowly she began to walk, trying to find her way out of the unknown field she had suddenly found herself in. Her mind was spinning and despite the warm summer air, she felt cold, almost frozen. The young woman stumbled in the dark for almost ten minutes before she finally came to a gravel road. As far as she could tell, this was not the road she had previously been on, as it had been two laned and paved.

As she walked down the quiet country road, Laurie Nelson finally came across a lone house with its porch light on. Fortunately, the elderly couple inside answered her frantic knocks and brought her inside and called the Sheriff.

When the Sheriff arrived, Laurie realized that she was not where she thought she should be. Before, she had been east of Ames in Story County. Now, unbelievably, she was told that she was in Boone County, west of Ames and more than fifty miles from where she last remembered being.

Later that evening Laurie's car was found abandoned on the road she had remembered driving on. The engine was still running, its lights were on, and all the doors were locked. It was as if the driver had completely vanished. But Laurie had not completely vanished. Instead, in the blink of an eye, she ended up miles away, alone in the middle of a dark field with no knowledge of how she got there.

Nor did Laurie Nelson ever discover what had happened to her on that warm August evening. The part of the road where she believes she disappeared has nothing to distinguish itself from the miles of asphalt through the lonely countryside. She suffered no ill effects from her ordeal. There were no physical injuries or bad dreams afterwards. Laurie was left with only the mental pain of knowing she had been through a completely mysterious event that defied all logic, and had no hope of explaining how she was able to travel more than fifty miles in less then two minutes.

What happened to Laurie Nelson that night in 1985? Investigators have suggested that she suffered from some sort of UFO phenomenon and experienced "missing time." However, instead of losing time, Laurie seemed to have lost space. She appeared to have been transported almost instantly by some unknown means through space to a different location. Was this an accident, or was there some intelligence at work?

The Mysteries of Teleportation

What happened to Laurie Nelson was by no means a unique incident. History is full of similar strange happenings that seem to defy the laws of time and space. Charles Fort in his book *LO* came up with the term teleportation to describe people and things that appear to have been transported distances by other then normal means. "If there have ever been instances of teleportations of human beings from somewhere else to this Earth, an examination of infirmaries and workhouses and asylums might lead to some marvelous disclosures. Early in the year 1928, a man did appear in a town in New Jersey, and did say that he had come from the planet Mars. Wherever he came from, everybody knows where he went after telling that."

There have been many reports of visitors from elsewhere mysteriously appearing from time to time. In 1954, the Japanese authorities detained a man trying to enter the country with a passport that revealed he was from an unheard of country named Taured. A thorough check was made by the customs officials to see if there was such a place anywhere on Earth, but they drew a blank. The stranger refused to throw light on the whereabouts of the mysterious nation of Taured and quickly left Japan.

A similar incident occurred in 1851 when a man calling himself Joseph Vorin was found wandering in the German village of Frankfurt-an-der Oder. When the German authorities asked the man where he was from, Vorin told them that he was from Laxaria, a country on the continent of Sakria. This baffled the authorities because neither of the places existed anywhere on their map of the world.

In 1905, a young man who was arrested in Paris for stealing a loaf was found to speak an unknown language, and after a lengthy interrogation session, the man managed to convey that he was from a place called Lizbia.

Thinking he meant Lisbon, the man was shown a map of Portugal, and a Portugese interpreter was brought in to talk to the young offender, but it was soon established that the man was not from Lisbon. The language the youth spoke was not an invented babble either; it had all the consistent syntactical rules of a language similar to Esperanto. Eventually, the strange-speaking man was released - never to be seen again.

Were these men all unwitting victims of some kind of strange phenomenon that teleported them from their homes, and into a strange, to them, place called Earth? Even more unsettling, have people from earth been caught up in some kind of hole in time and space and teleported far beyond this planet or even reality?

Research shows that there are old German folklores about just such phenomena. Of course these stories are filled with the beliefs of the day, attributing such strange disappearances to ghosts and spirits. Called the Wilde Jagd, the "Wild Hunt" were a group of ghosts or demons in the shape of ugly hunters and horsemen who flew together with other devils and demonic animals through the air to catch people on earth and take them into the sky for eternity. This sounds at first absolutely unbelievable and like pure superstition, but if we have a closer look at some of the old tales, the whole weird story about the "Wild Hunt" reminds us of modern tales of teleportation and strange disappearances.

Gordon Creighton, writing in his article: "*A Brief Account of the True Nature of the UFO Entities*," for **Flying Saucer Review**, states that ancient Islamic tales of supernatural beings called Jinns would often tell of Jinns "snatching up humans and teleporting or transporting them, setting them down again – if indeed they ever do set them down again, miles away from where they were picked up, and all this in the twinkling of an eye."

Teleportation: From Star Trek To Tesla

Manmade Or Natural Forces?

Some science fiction movies, television shows and novels use teleportation in their storylines as an integral part of their plots. Usually teleportation is achieved in science fiction by the use of some exotic, futuristic technology that enables the characters to transport themselves great distances in the blink of an eye. Does this mean that we have to wait for actual science to catch up with the imagination of writers when it comes to teleportation. Possibly not.

There is a good chance that teleportation of some kind does take place on a regular basis, using natural laws that are still, little understood today. A good example of "non-technology" teleportation occurs during a particular form of haunting.

There is extensive literature which describes what are known as poltergeist phenomena. Psychic researchers have sifted through firsthand accounts, judging them by the same criteria as those applied to historical, anthropological and forensic source material. Other psychic researchers have themselves been fortunate enough to observe phenomena, and have written careful or less careful accounts.

On this basis most scholars have concluded that many strange physical phenomena have really occurred at intervals throughout history. The thinking public is less inclined to be skeptical about these events than it is about other psychic phenomena; there is a sense of "it doesn't happen to me, but all the same it seems to happen to some people."

The modern technical term for physical poltergeist phenomena is RSPK (recurrent spontaneous psychokinesis). In a poltergeist haunting, objects sometimes spontaneously fly about the house, apparently in a random way, without discrimination. But it is seldom that anyone is hurt by them.

Sometimes their paths of flight are unnaturally crooked, and often the objects are not seen to leave their normal positions; sometimes they just appear in the air and gently drop. Sometimes they arrive warm and sometimes they arrive with spin, or angular momentum. Sometimes they appear in motion, their flight starting from a position different from their normal place.

Sometimes the origin of the objects is unknown, as in the cases where showers of stones are reported. Furniture, heavy and light, can move spontaneously about the room, tip over, and even levitate and crash down again on the floor. Sometimes these movements are observed, sometimes they happen when no one is in the room, for example, during the night.

A typical poltergeist case will last only for a period of weeks or months. The spontaneity of the movements has led to the opinion that the phenomena are caused by a ghost. However, the more usually accepted view is that one or more of the personalities involved, usually children, are unconsciously responsible for the phenomena.

Teleportation: From Star Trek To Tesla

Such personalities have been called epicenters (although this term is also used to describe the area of the house in which events most frequently happen). When the subject realizes that he or she is responsible, and is thereby the center of attention, he often adds to the effects by normal physical means.

Rarer and stranger events, including many quasi-physical as opposed to physical phenomena, have been reported in poltergeist cases, but the particular type of event that has been most often observed is the traveling of an object from one location to another in an abnormal way. It might best be described as the disappearance of the object in its original position, and its re-appearance somewhere else. In other words, teleportation.

Writer Cynthia Sue Larson calls such strange events Reality Shifts: "The manifestation of objects appearing, disappearing, transforming and transporting. Changes in the way we experience time. Any sudden, abrupt alteration of physical reality with no apparent physical cause. The source of synchronicity."

Shifts in reality have long been known by those who practice magick, work as shamanic or energy healers, or study unusual phenomenon. These shifts are now becoming known to the general public as a very real phenomenon that affects us all. And without the help from a mechanical device.

Something very exciting has happened for the future of the unification of science and metaphysics. This new evidence could be the greatest step forward towards a scientific realization that there might be some possibility for the existence of psychism, magic and miracles. What has happened is something called quantum teleportation.

Quantum teleportation is a somewhat mind boggling concept, but it's really not all that hard to explain philosophically. The mathematics are extremely complex and does something very unique in the annals of science. It shows that it is possible for one thing to effect another without any intervening mechanism involved.

This has been the one stumbling block that has always been tossed by scientists towards believers in things like psychic abilities, magick, miracles, astrology, etc. Scientists have always pointed to their scientific theories on action and reaction, cause and effect, particles and waves.

They say things such as: "If mysticism exists – then show us the mechanism involved. Show me the particles or waves or signals that cause this communication across vast distances at nearly infinite speed."

Now there appears to be just such proof. Scientists at IBM and in Europe have shown that action at a distance can happen with no mechanism at all. Hard as it may be to believe. This is true science. In fact, the implication is that this action may happen at infinite speed.

Of course this idea is nothing new to anyone who has taken the time to research unusual phenomena. Thousands of books have been written over the years, filled with anecdotal tales of mysterious events that can't be explained by modern science. At least until now!

Teleportation: From Star Trek To Tesla

Appearing And Disappearing

Simon Harvey-Wilson wrote in the ***The Journal of Alternative Realities*** (Volume 5, Issue 1, 1997) that throughout history there have been reports from paranormal research, shamanism, mysticism and UFOlogy, of people or objects that become invisible, materialized, dematerialized, or teleported.

Because of their similarity, understanding the dynamics of one of these phenomena may assist us in understanding the others. In other words, perhaps the physics, if that is the appropriate term, of becoming invisible may be similar to that of the materialization and dematerialization of solid objects, which in turn might be similar to that of teleportation.

Materialization and dematerialization are opposite sides of the same coin, and sometimes would be indistinguishable from invisibility. But, if someone simply disappears from a witness' sight, later reappears, and could not be touched while invisible, we can assume that something other than an inhibition in the witness' perceptual system has occurred.

In the annals of the paranormal there are probably more instances of things materializing than dematerializing. In seances for example, objects have frequently been known to appear, seemingly out of thin air.

Called apports, it is generally assumed that a disembodied spirit has either created them out of nothing, or teleported them from elsewhere. An example might be a fragrant rose, still covered in dew, that suddenly falls out of the air onto a seance table. If such an apport was somehow picked off someone's rose bush by a spirit, one wonders what the owner of that bush might have seen if they were looking out the window at the time.

The seance room is not the only source of mysterious apports. Areas reported to be haunted by ghosts or poltergeists are often plagued by the appearances and disappearances of everyday objects. The English healer Mathew Manning, who experienced a lot of poltergeist activity during his teenage years, gives several examples in his book ***The Link***.

One time I was out collecting material for a Guy Fawkes fire at the bottom of our garden. Finding myself well short of rubbish, except for half a dozen cardboard boxes, I went to the house and asked my mother what else I could use. There was no one else at home and she had no good idea or suggestion. I returned to the bottom of the garden, and to my utter amazement I found a stack of large logs and wood placed next to the cardboard boxes. At that time there was nobody who could have done this, let alone in the short space of time I had been in the house. In all, there were several hundredweight of wood and logs that had seemingly appeared out of nowhere. Other such apports included several gramophone records, a bag of sugar, a bank note, a pair of black lace gloves and postage stamps.

Teleportation: From Star Trek To Tesla

In another incident, "a pint bottle of beer and an apple pie..." appeared in his bag while he was on a train. Manning also describes apports that seem to have come from somewhere else. "A long-playing record of which I had a copy appeared one day in the house; it seemed to have come from another owner as it bore obvious marks of wear. There seemed no reason for this to materialize as I owned a copy of it already."

British writer Colin Wilson recalls a poltergeist related teleportation incident in which an egg, apparently from the kitchen refrigerator, floated in through the lounge room door of a poltergeist affected house, and dropped onto the floor.

One of the house occupants then put all the refrigerator eggs into a box and sat on the lid. As if provoked by this defiance, eggs continued to smash all over the floor until the box was empty, despite its remaining closed throughout the event.

It has been traditional to regard poltergeist activity as the pranks of invisible spirits or curses with black magick. More recently it has been suggested that the phenomenon may be linked to unresolved conflicts in the mind of a teenager, or emotionally upset individual living in such a house: a form of unconscious psychokinesis working through hyperspace.

This theory makes several radical assumptions about the nature of consciousness: for example, that it can affect matter at a distance. Poltergeist-like events also occur after UFO contact and abduction cases.

One suggestion is that some UFOs may be able to teleport through "hyperspace," which is in turn somehow connected to consciousness, so that being pulled into this realm affects abductees' minds deeply enough to cause poltergeist activity around them afterwards.

I dries Shah, an expert on Sufism, which is the mystical branch of Islam, claims that the Qutub, the chief of the Sufi system, is always someone who has attained the degree of Wasl (Union with the Infinite). Such men "are able to transport themselves anywhere instantaneously, in physical form, by a process of decorporealization."

This sounds like teleportation, and reinforces the claim that such abilities are linked with altered states of consciousness. The late parapsychologist, D. Scott Rogo, pointed out in his book, **The Haunted Universe**, that teleportation overlaps the phenomenon of bilocation, whereby a person is seen in two places at once. The Italian monk Padre Pio apparently appeared physically in two places simultaneously on several occasions.

John Michell gives an example of apparent teleportation in his book **The Flying Saucer Vision**. On 25th October 1593, a Spanish soldier was arrested in the main square of Mexico City because he was unable to account for his presence there, and because he was wearing the uniform of a regiment that was at that time stationed in the Philippine Islands, nearly a year's travel away by ship.

The befuddled soldier nevertheless gave precise details of his life in Manila up to the moment he had found himself instantaneously and inexplicably transported to Mexico. He was even able to tell his interrogators of the recent death of the Spanish governor of the

Teleportation: From Star Trek To Tesla

Philippines; news that finally vindicated the Spanish solider did not arrive in Mexico City for many months.

It is interesting to wonder what could have caused this event. Did the soldier possess unknown psychic abilities, was he unusually devout, or was he perhaps in the wrong place at the wrong time when some delinquent spirits or aliens decided to have some fun at his expense? Or could he have been swept up in some kind of window or door through time and space? This is an event that could happen naturally when the conditions are right, and some poor soul accidently stumbles through. Possibly to be teleported someplace else on the planet, or to vanish forever to some unknown land.

The parapsychologist Professor Erlendur Haraldsson quotes various witnesses who, with other devotees, in the late nineteen-forties, used to go for afternoon walks with the Indian religious leader Sathya Sai Baba towards the river in his home village of Puttaparti, in Southern India.

On several occasions Sai Baba would disappear from among the devotees and reappear at the top of a nearby hill. Sometimes he would then shout that he was coming down and would instantly reappear among the devotees.

Later, in 1995, there were anecdotal reports that in full view of a group of Australian devotees, who had been granted an interview with him at his ashram at Whitefield on the outskirts of Bangalore, Sai Baba teleported an elderly man back to his home in Australia to be with his ailing wife. His friends saw the man disappear from the interview room, and when, just after the interview, they went and phoned his home in Australia, it is claimed that it was he who answered the phone.

If true, this report suggests that someone with powerful paranormal powers can teleport another person with only the power of his mind.

In literature on shamanism there are instances where objects, or even living insects, are materialized. In his book, *Gifts of Unknown Things*, Lyall Watson describes an incident in the Amazon where he witnessed a local healer first remove an infected tooth from a patient, and then announce that he had to make the pain of the infected gum go away.

To do this, he somehow materialized over a hundred black army ants which marched in an ordered column out of the patient's mouth, down his arm and away into the grass at the edge of the clearing. This caused great mirth among the watching natives because, as Watson later discovered, the local word for pain was the same as that for army ant.

As Watson put it: "The healer had promised that the pain would leave, and so it did, in the form of an elaborate and extraordinary pun. It walked out."

A brief article in *New Dawn* (July-August 1997) claims that a U.S. Defense Intelligence Agency translation of an article in a 1983 Chinese journal described successful experiments on the teleportation of small objects such as fruit flies, a watch, a match and a nail using "extraordinary children" as test subjects.

Teleportation: From Star Trek To Tesla

This research was documented by Zhu Yi Yi, of Shanghai, a biology graduate of the Shanghai Fu Dan University, and staff writer for *Ziran Zazhi* (Nature Journal). A translation and summary of her work is found in *Incredible Tales of the Paranormal* edited by Alexander Imich, Ph.D.

The first case brought to Zhu Yi Yi's attention was that of a young girl named Yang Li who could "remove the cigarettes." Ms. Zhu herself counted the cigarettes which were put inside a cardboard box with a lid. After awhile, Yang Li said "one had been removed." The box was opened, the cigarettes counted, and sure enough, one was missing. Where it went to is unknown.

There were also children who could teleport something into a closed container. In this case, the containers were teacups with lids on them and the objects teleported into the cups were flowers and flower buds. The source of one of the Jasmine buds was discovered. It came from a Jasmine plant growing in a pot on Mr. Yang's balcony.

This incident took place at the home of Yang Li, mentioned above (in Chinese the Surname comes first) in the town of Kun Ming, famous for its flowers. Zhu Yi Yi invited two 12 year old girls trained for two years under Professors Luo Xinfun and Zheng Tianming, who also came along for the test. There were four young girls in all, and all were able to teleport flowers and buds into the teacups, although none of them had tried this particular experiment before.

The researchers concluded that: "Transference is not a simple process of mechanical movement in three dimensional space."

If such reports are true, we might suspect that the original Chinese article prompted the U.S. military to sponsor similar research. Dr Richard Boylan claims that researchers at the Lawrence Livermore and Sandia National Laboratories in the U.S. have conducted "successful teleportation experiments." Not surprisingly details do not seem to have been published in any science journals, so it is hard to know what to make of such claims.

How could something dematerialize and/or teleport from one place to another? A common explanation is that these objects enter other dimensions invisible to normal human perception.

Unfortunately this is not a very satisfactory explanation because it simply replaces one mystery with another. Nevertheless, the possible existence of higher dimensions, otherwise known as hyperspace, is frequently mentioned by physicists these days. The advantage of hyperspace is that, being beyond the three physical dimensions of space/time, it may facilitate shortcuts from one part of time space to another.

Topologists, who study other dimensions from a mathematical perspective, point out that three dimensional physical barriers, such as the sides of a box, cease to be obstacles in higher dimensional space. In 1985 the U.S. physicist Kip Thorne suggested that inter-dimensional shortcuts called wormholes might one day facilitate space travel.

Teleportation: From Star Trek To Tesla

However, physicists claim that to create a wormhole would require vast amounts of energy. It is even suggested that black holes are versions of such rips in the fabric of time and space.

On a much smaller, but no less dramatic, scale, perhaps consciousness itself is somehow able to create the equivalent of a wormhole to facilitate teleportation. If so, then perhaps an advanced extraterrestrial civilization has researched this aspect of the physics of consciousness enough to use the results in the UFO and abduction phenomena.

In his book *Hyperspace*, Professor Michio Kaku describes how the Superstring theory postulates that numerous other dimensions exist beneath the sub-atomic scale, and that electromagnetism and the other three fundamental forces in the universe are united in this realm.

Perhaps the matter produced from the energy of the Big Bang, and that produced by anyone that materializes objects, originates within hyperspace to which consciousness also has access. In fact, within hyperspace, matter, energy, the fabric of time and space, and consciousness itself, may derive from the same source.

Another explanation for teleportation is that the object concerned dematerializes, somehow travels to its destination, and then becomes solid again. In his book *The Physics of Star Trek*, the physicist Professor Lawrence Krauss discusses the scientific validity of the science fiction ideas in that popular TV series.

Krauss points out that, from a physicist's perspective, to teleport a human body, as in "Beam me up Scotty," requires several steps. Firstly you have to record the exact configuration of all the atoms in the body, and to store that much information would require an astronomically tall heap of 10-gigabyte hard drives.

Secondly, you would need to somehow dematerialize the person, which he claims requires vast amounts of energy. Thirdly, you transmit to the new location either the body's sub-atomic particles, called quarks, or perhaps just the atomic information about them.

Finally, either with the original quarks or some new ones, you use the information about the person's body to rematerialize it at the other end. Krauss is wise enough to add the disclaimer that if humans have souls, as many people believe, his whole plan falls to pieces. Nevertheless, given these and several others obstacles, Krauss is of the opinion that science won't be teleporting anyone anywhere for some time to come.

We will examine in later chapters other possible methods of teleportation that could circumnavigate the technological problem of taking something apart bit by bit. Considering the strange stories throughout history that could be examples of teleportation by advanced mental abilities, spirits and poltergeists, natural environmental events etc, it must be argued that time and space is a lot more pliable then is thought by current science. If this is true, then workable means of teleportation could be a reality a lot sooner then we think. You might even learn to practice a form of it yourself!

CHAPTER 2
SCIENCE FICTION - SCIENCE FACT

Is it possible that one day we will be able to enter a teleportation booth, have our atoms taken apart, transmitted to a receiver booth some distance away and reassembled no worse for the wear? Until now, such ideas seemed to be strictly within the realm of science fiction with Captain Kirk "beaming" with the greatest of ease all over the galaxy.

Now, however, it turns out that in the strange world of quantum physics, teleportation is not only theoretically possible, it can actually happen.

One group of researchers at the University of Innsbruck in Austria published an account of the first experiment to verify quantum teleportation in the December 11, 1997 issue of *Nature*. And another team headed by Francesco De Martini in Rome has submitted similar evidence to *Physical Review Letters* for publication.

Neither group has managed to beam themselves to the moon, yet what they did prove is still pretty amazing. Anton Zeilinger, De Martini and their colleagues demonstrated independently that it is possible to transfer the properties of one quantum particle (such as a photon) to another – even if the two are at opposite ends of the galaxy.

Until recently, physicists had all but ruled out teleportation, in essence because all particles behave simultaneously like particles and like waves. The trick was this: they presumed that to produce an exact duplicate of any one particle, you would first have to determine both its particle-like properties, such as its position, and its wavelike properties, such as its momentum.

Doing so, however, would violate the Heisenberg uncertainty principle of quantum mechanics. Under that principle, it is impossible to ever measure wave and particle properties at the same time. The more you learn about one set of characteristics, the less you can say about the other with any real certainty.

Einstein originally proposed the possibility of quantum teleportation. However he presented it as a paradox that was meant to disprove quantum mechanics. This supposed paradox was called "The Einstein-Podlosky-Rosen (EPR) Paradox."

The usual view of quantum mechanics says that a wave function determines the probabilities of an actual experimental result and that it is the most complete possible specification of the quantum state. Einstein believed the predictions of quantum mechanics to be correct, but only as the result of statistical distributions of other unknown but real properties of the particles.

In 1993, though, an international team of six scientists proposed a way to make an end-run around the uncertainty principle. Their solution was based on the Einstein-Podolsky-Rosen effect theorem of quantum mechanics.

Teleportation: From Star Trek To Tesla

When two particles come into contact with one another, they can become entangled. In an entangled state, both particles remain part of the same quantum system so that whatever you do to one of them affects the other one in a predictable, domino-like fashion. Thus, the group showed how, in principle, entangled particles might serve as transporters of sorts. By introducing a third message particle to one of the entangled particles, one could transfer its properties to the other one, without ever measuring those properties.

These ideas were not verified experimentally until the Innsbruck investigators performed their recent experiment. The researchers produced pairs of entangled photons and showed they could transfer the polarization state from one photon to another.

David Bohm is a physicist who studied under Neils Bohr and has written many renowned books on the subject of quantum mechanics. He also taught at Princeton where he had discussions with Einstein. Like Einstein he wasn't very happy with some of the implications of quantum mechanics. For instance the idea that particles have no existence until they are observed by someone.

Bohm's answer to these problems was to come up with a new theory that postulated something called a **Quantum potential** which is a kind of weak type nuclear force which does not diminish with distance. Using this model he was able to explain all of the unusual effects of quantum mechanics and Einsteinian theory. Eventually it occurred to him that what he was proposing was that on the sub nuclear level of existence, there is no such thing as distance or space. It's all here and now.

This is one of the only ways that Bohm could explain such seeming paradoxes as single electrons which can exist in several locations at the same time, or particles that can split into two negative and positive particles and still have the proper polarity when viewed simultaneously by two different people (IE: Quantum Teleportation).

In 1951, Bohm presented a paper in which he described a modified form of the Einstein-Podolsky-Rosen thought experiment which he believed to be conceptually equivalent to that suggested by Einstein. (1935), but which was easier to treat mathematically.

Bohm suggested using two atoms with a known total spin of zero, separated in a way that the spin of each atom points in a direction exactly opposite to that of the other. In this situation, the angular momentum of one particle can be measured indirectly by measuring the corresponding vector of the other particle.

In this model of reality, space and the dimensions are not much more than an illusion of perspective created in the mind. It is created in the same way that a three dimensional illusion is created by a hologram.

When we look at the universe we only see what looks like three dimensional space, perhaps because our perspective is too limited to see the actual reality. In other words Bohm's theories more or less state that the universe is a big hologram.

Teleportation: From Star Trek To Tesla

Quantum Teleportation

In quantum teleportation, two objects are first brought into contact and then separated. One object is taken to the sending station, while the other is taken to the receiving station. At the sending station, the second object is scanned together with the original object which one wishes to teleport, yielding some information and totally disrupting the state of both objects. The scanned information is sent to the receiving station, where it is used to select one of several treatments to be applied to the third object, thereby turning the third object into an exact replica of the first object.

The dream of teleportation is to be able to travel by simply reappearing at some distant location. It might appear that one could scan the object and send the information so that the object can be reconstructed at the destination. Yet, fundamental laws like the Heisenberg uncertainty relation do not allow one to measure any object to arbitrary precision. Scientists have suggested that it is possible to transfer even quantum states, provided one does not get any information about the state. This becomes possible by utilizing entanglement, one of the essential features of quantum mechanics.

By producing pairs of entangled photons by the process of parametric down-conversion and using two-photon interferometry for analyzing the entanglement, we could transfer a quantum property (in our case the polarization state) from one photon to another.

In the quantum teleportation process, physicists take a photon (or any other quantum-scale particle), transfer its properties (such as its polarization, the direction in which its electric field vibrates) to another photon, even if the two photons are at remote locations. What's important to emphasize is that this scheme doesn't allow physicists to teleport the photon itself – only its properties to another, remote photon.

Of course this scheme is intended only for quantum-scale particles, such as photons and atoms. Although no existing laws of physics prevent quantum teleportation from being carried out with living humans, it is extremely unlikely that this scheme could be carried out in such a large scale. This is because the uniquely quantum properties, such as entanglement that make teleportation possible, quickly break down as objects scale up to the macroscopic sizes.

Defense News, a defense industry trade publication, reports that the Pentagon is well on the way to developing the ability to "teleport messages across the battlefield and around the globe" faster than the speed of light. The Pentagon, in cooperation with the U.S. Army Research Office, the National Security Agency and the Advanced Research Projects Agency, are well on the way to implementing such technology within the next decade.

Henry Everitt, a physicist at the Army Research Office in Durham, North Carolina, explains that the concept is based on a portion of quantum physics theory involving photons.

Teleportation: From Star Trek To Tesla

Photons are units of electromagnetic energy that carry momentum and energy yet have no mass. Without mass, photons are not bound to the speed of light, usually thought of as nature's ultimate speed limit. Teleporting communication has an advantage over conventional methods in that teleported messages are more secure and don't give away the location of either the sender or the receiver.

Everitt says, "This has gone from nothing – I mean no money at all in the world – to being one of the hottest areas in physics, one of the hottest areas in science, and one of the largest programs we have here at the Army Research Office."

What would these faster-than-light communicators look like? Everett says they would look pretty much like a standard laptop or even a cell phone, but the inner workings would be very different.

Though tight-lipped about the details, Everitt told *Defense News* that a major contract will be awarded soon and that other business opportunities will arise as the technology moves out of the laboratory into the real world. "It has been demonstrated that information can be teleported over 40 kilometers using existing technology," he explains.

But those experiments are just baby steps compared to their quantum potential. "There is no limit to the distance over which information can be teleported."

As advanced as faster-than-light communication sounds, it's only one segment of a much larger project, the Quantum computer. Everitt says: "The Holy Grail of all this is to build a quantum computer that will do things a classical computer will never do." The unique characteristics of the photons to exist in multiple states and not be bound by the speed of light means that quantum computers will be able to form many different calculations at the same time, far outpacing today's computers.

The Quantum Information Science Program reportedly involves thirty-four separate projects by researchers at twenty-one universities, two corporate laboratories and three of the nation's most highly secretive government agencies.

Scientists hope that quantum computers will be infinitely faster and more powerful than present-day computers. Caltech physics professor Jeff Kimble, thinks that we are just now seeing the development of a whole new way of thinking.

"I believe that quantum information is going to be really important for our society," Kimble says. "Not in five years or ten years, but if we look into the 100-year time frame it's hard to imagine that advanced societies don't use quantum information."

And in principle, teleportation could be used to send information to create replicas of objects, not just light beams. Researchers are already looking to teleport atoms. Could this mean the transporters of Star Trek could one day be a reality? "I don't think anybody knows the answer," Kimble said. "Let's teleport the smallest bacterium first. How much entanglement would we need to teleport such a thing?" Would such a teleported bacterium actually be the same bacterium, or just a very good copy? No one knows for sure.

Teleportation: From Star Trek To Tesla

Call It Science Fiction

In a clever essay entitled: ***Teleportation: or the penetration of the wave function into a classically forbidden region***, author Ian Woolf uses the ideas put forth by various science fiction writers to speculate on the realities of science-based teleportation.

Woolf notes that from Harry Harrison's matter transmitters to Alfred Bester's Jaunts, teleportation has long been a favorite in science fiction stories. Some authors, like Larry Niven, take great care to try and get all the scientific implications right, while others like Star Trek misunderstand them.

The traditional Newtonian method is matter transmission or "beaming." The matter transmission has usually been visualized as a breaking up of the original substance by scanning the molecules and atoms one at a time, then blasting out this scanning signal to be rebuilt at the receiving end.

In Poul Anderson's ***The Enemy Stars***, a side effect is that the body is vaporized, so that one winds up with a complete record of the passenger plus a cloud of superheated plasma. The gas is sucked through a grid to be stored in a matter reserve, to await the next incoming signal. The record of the passenger is sent across space by radio. A receiver picks it up and uses it, plus the plasma in its matter reserve to reconstruct the passenger. As Niven says, "I wouldn't ride in one of the goddamned things."

The earliest recorded story of a matter transmitter was in Edward Page Mitchell's 1877 novel, ***The Man Without a Body***. In his book, a scientist invents a machine that breaks down the atoms of a cat and transmits them by wire to the receiver, where the animal is reassembled alive and well. He tries it upon himself, but the battery dries up before he can transmit more than just his head.

A recurring theme in many science fiction stories about teleportation is the question of whether or not a person going through a teleportation chamber, is the same person that reemerges at the final destination. The idea that human consciousness and personality could be downloaded into a very advanced computer or other organic form has become a popular element in many stories in recent years. It is often introduced as a means for humanity to achieve immortality by escaping our ageing and vulnerable bodies and is therefore generally popular in the fictional societies that have access to the technology.

Interestingly, many stories contain arguments against the process as destroying the essential nature of human personality – for example the Adamists' beliefs in Peter Hamilton's ***Night's Dawn Trilogy***. Such technology would rely on the premise that the essence of human self-awareness and personality is solely based on data stored in the molecular structure of the brain and would discount the presence of any non-physical spirit. Many people however do believe in such a spirit and those who have doubts may well be unwilling to risk their immortal souls by undergoing the process – especially as it often involves the destruction of the human brain in the procedure.

Is the downloaded person the same as the original, a self-aware copy or just a programmed response routine that mimics the original? These dilemmas when faced with the possibility of near immortality are explored in a number of stories.

In the short story **Ginungagap** by Michael Swanwick, aliens offer humanity the chance to participate in their civilization by sending an individual through a black hole across light years to the aliens' world. In the process, each particle of matter sent through the black hole disappears and another appears on the other side from which a body is reconstructed.

Much of the story is taken up with the female volunteer trying to decide if the person who would emerge on the other side of the trip would really be her or not. She finally decides to take the trip and the story ends with the aliens making many copies of her with her atomic data when she reaches their world.

Which of them, if any, is really her? The implication is that none of them are and that she died in the data-gathering process. There is an implication here for any teleportation device such as the transporter beam in Star Trek. This is explored in one episode when it is discovered that the transporter has made a copy of Riker. It is impossible to decide which is the original and both are treated as equally Riker.

However quantum physics suggests that there is a good explanation for the destruction of the original. Measuring a quantum system changes what you are measuring. The act of observation will change the original state to something unpredictable, so you'll have your information, but the original state of the brain has been scrambled.

There are only two distinct physical states of matter, definite-observed and uncertain-unobserved, what Greg Egan in **Quarantine** calls "smeared." Making an observation of a smeared electron will change it. There are many qualities that are incompatible, the measurement of one fuzzes the other further into uncertainty.

Quantum logic, as might be used in human brains and future computers, relies on the smeared state for its computational power. Thus accurate copying of a complex quantum system is impossible, and the attempt will change the system you are attempting to copy.

Such stories do not need to invoke a non-physical spirit to make their point but merely try to demonstrate that the duplication or transfer of personality data produces a copy and not a transfer of the original individual. The idea of downloaded consciousness resulting in immortality for the individual is thus exposed as an illusion.

Personal self-consciousness remains a mystery and its subjective nature as an experience sets it outside of completely objective scientific study. It would seem to be always impossible to be sure if any conscious experience had been adequately duplicated. In the same way it is not possible to be sure that any two individuals even have the same experience of what they both call the color red, let alone more abstract concepts such as love.

Teleportation: From Star Trek To Tesla

Looking Beyond Science

In his book, *The Physics of Star Trek*, the physicist Professor Lawrence Krauss says that the volume of atomic data about the human body is unmanageable. However, the mathematics of Fractal Geometry, apart from producing beautiful psychedelic patterns, enables redundant data to be removed from, for example, a high resolution spy satellite image, to facilitate its transmission back to earth, where, using the same mathematics in reverse, the image can be decompressed.

The final product is of high quality, and such techniques are rapidly increasing in sophistication. Fractal Geometry can also produce patterns that are similar to many of those found in nature. As the human body is made of trillions of almost identical sub atomic particles, data compression would assist in transmitting such information. So Krauss' information overload objection is probably irrelevant.

Krauss claims that to vaporize the body into pure energy in preparation for teleportation would take the equivalent of a thousand 100-megaton bombs. Yet paranormal reports of teleportation do not mention such energies. We could suggest therefore that modern physics is investigating the nature of matter the hard way – from the outside. Paranormal evidence suggests that the subtlety of consciousness can affect matter from the inside in a very energy efficient manner.

A simple example of this is paranormal spoon bending, where the mind seems able to affect the molecular structure of metal from within. At present we don't know how this works, but we will never find out unless we do the relevant research. One place to look is the relationship between matter, energy, consciousness, and the domain in which they operate, called space/time, which is increasingly being seen by physicists and others, not as the emptiness in which things happen, but rather as a substance that can expand, contract, bend or reverberate.

This suggests that space/time may have an outside or beyond, which might just be an alternative description for hyperspace. Attempts to grapple with a definition of this beyond have referred to altered states of consciousness, other wavelengths, vibrations, or, more esoterically, some sort of transcendental consciousness or universal mind.

Krauss' suggestion, that in teleportation we may only need to transmit atomic information rather than atomic particles, echoes this idea; that information theory may provide the best model for the fundamental nature of reality. In other words, the basic units of matter, if such things exist, may be units of information rather than anything solid.

As an example of seeing beyond space/time, physicists claim that no one needed to show the energy produced during the Big Bang how to coalesce into matter. Sub-atomic particles and atoms seemed to know how to assemble themselves, as if the rules of physics were

Teleportation: From Star Trek To Tesla

already there. If this is so, how was this information stored, and where did the energy of the Big Bang actually come from?

Some researchers speak of the energy of the vacuum, or Zero Point Energy, which suggests that behind the fabric of space/time there may exist an almost infinite amount of energy. For example: "According to quantum theory, empty space is not as empty as it seems: if we could examine a vacuum at the Plank scale – a resolution of 10-35 meters – we would see a seething mass of virtual particles, including photons, flitting in and out of existence." (Lyall Watson, 1996)

If the entire universe popped up from nowhere, it does seem rather childish for physicists to claim that it is impossible for someone with powerful paranormal ability, such as Uri Geller, to produce an object containing less than one kilogram of matter from that same nowhere. What we'd like to know is how he does it. Perhaps consciousness can somehow address the energy of the Plank scale, and persuade it to create permanent atomic particles rather than just virtual ones.

But how would these particles know what object to make? To answer this we need to refer back to Krauss' earlier disclaimer that, if humans have souls, his teleportation theories are probably wrong. However, the existence of souls might make teleportation easier to explain rather than harder.

There is evidence from events such as Near-Death-Experiences that at least some humans do have invisible forms of consciousness that could perhaps be called souls. It has also long been claimed that all living things have something resembling a subtle version of their genetic code that exists beyond the body.

Plato referred to the Realm of Forms, and in modern times Rupert Sheldrake speaks of Morphic Fields. He suggests that as things grow they obtain developmental information from both their genetic codes and Morphic Resonance. These two informational sources may even overlap, with one able to substitute for the other.

Along these lines, a brief, article in *Nexus* magazine (Kanzhen, 1995) claims that a Chinese scientist working in Russia has perfected a bio-electromagnetic field process whereby he can change the genetic structure of some plants and animals. If true, such a discovery would be of enormous importance. This might mean that the information needed to reassemble the human body after teleportation is obtainable from an informational realm like Morphic Resonance.

Sheldrake has suggested that Morphic fields may transcend time and space, which might mean that, provided the body was disassembled correctly, the information to reassemble it would not need to be transmitted anywhere, but instead might be stored non-locally and therefore be accessible from anywhere.

After several landmark experiments, modern physics has accepted that non-locality does exist at the sub-atomic level. Otherwise known as the Holographic Paradigm, this research

27

Teleportation: From Star Trek To Tesla

shows that within the quantum realm, something that occurs in region A can have an instantaneous physical effect in region B, regardless of the distance or conditions between A and B. Further research in this field may yet lead to significant advances in our understanding of both the paranormal and the nature of consciousness.

In 1989, Professor Roger Penrose at Oxford University put forward the controversial suggestion that consciousness may have something to do with the quantum realm. Obviously more research is needed, but if he and other theorists in this subject are correct, then perhaps consciousness has the capacity to reach beyond the dimensional limitations of space/time, to an astonishingly creative nonlocal informational realm which holds something like the blueprints for the structure of matter.

Once accessed, willpower alone may be able to solidify such information to produce, within space/time, something that the inhabitants of that realm normally regard as solid. These suggestions imply that scientists are going to have to take a harder look at the evidence for paranormal anomalies such as teleportation, materialization, invisibility, and the UFO phenomenon if they are ever going to discover the fundamental nature of reality that they claim to seek.

It is also interesting to observe that a form of meditation meant for the awakening of Kundalini can also cause teleportation. When entering Samadhi, which is called the supreme meditation stage, a disciple feels great changes even to his material body. This is the very stage in which we get the experience of Five Divine Abilities. These include:

❑ Divine Legs, that make possible both levitation and teleportation.

❑ Divine Ears, that make possible listening to gods' voices and distant sounds.

❑ Divine Mind Reading, that allows getting information from other souls.

❑ Divine Knowledge of Past Lives, that makes it possible to remember the previous lives of another people.

❑ Divine Knowledge of Future Lives, which makes it possible to understand the place of your future reincarnation and to establish a connection with different realms.

These are the Five Divine Abilities that are described in Buddhist sutras. With this experience we can attain a view based on a true thorough knowledge. Therefore, a disciple can understand the wrong opinions he has. This is called a view based on a true thorough knowledge or the absolute freedom, absolute joy and absolute happiness that surpass the fundamental inconstancy. In Buddhist religion it is called Emancipation.

CHAPTER 3
BEAM ME UP SCOTTY!

No other television program has brought the idea of teleportation to the attention of the general public quite like *Star Trek* and its transporter technology. When *Star Trek* debuted in 1965, the idea that you could take a person apart atom by atom and "transmit" them to the surface of a planet from an orbiting spacecraft was a revolutionary concept.

The idea of teleportation was nothing new to those weaned on years of science fiction novels and magazines. Nevertheless, *Star Trek* was the first to show that you could land on a planet's surface without taking the entire spaceship out of orbit to a possibly messy and dangerous touchdown.

In the *Star Trek* mythos the transporter is considered the single greatest revolution in the movement of people and goods in recorded history; invented in 2205, this device cut trans-planetary transport times to near zero at a stroke.

The basic operating principles of the transporter are relatively simple. It makes a detailed scan of the subject, breaks down its molecular structure, then transmits this beam to another location. The information gained from the scan is then used to reassemble the subject exactly as before.

Like many simple ideas, the actual engineering required to construct a working transporter are quite more complex. A standard transporter unit consists of ten major components. The transport chamber is the area in which the subject is placed for transport.

The transport chamber can be of almost any size or shape, though larger chambers have far greater energy requirements and are correspondingly less efficient for general use. Most transport chambers are capable of holding approximately six persons.

The operator's console is the control unit of the whole system; these consoles are typically manned by a single operator who oversees the transport process and is responsible for reacting to emergency situations, as well as conducting routine maintenance of the transporter systems.

The transporter controller is a dedicated computer system which controls the minutiae of the transport process itself. In the time frame of *Star Trek* there have been almost no recorded transporter accidents for a number of years.

The primary energizing coils are located directly above the transport chamber. These coils generate the annular confinement beam, creating a space-time matrix within which the dematerializing process occurs. The primary energizing coils also generate a containment field to prevent any possible breach of the annular confinement beam during the transport process. This is important as such disruption can result in a sizeable energy discharge.

Teleportation: From Star Trek To Tesla

The phase transition coils are located in the floor of the transport chamber. It is the phase transition coils that cause the actual dematerialization/materialization process. They do this by decoupling the binding energy between the subatomic particles of the subject, causing the atoms themselves to disintegrate.

Molecular imaging scanners are located in the roof of the transport chamber. These devices scan the subject to be transported at quantum resolution, determining the location and momentum of every particle within the subject.

Bulk cargo can be scanned at the molecular resolution, as it is not generally vital to recreate the object exactly. Living matter requires that exact information be obtained, a process which violates the Heisenberg Uncertainty Principle. This is made possible by the Heisenberg Compensator System, a component of the molecular imaging scanners of all personnel transport systems.

All transporters are built with four redundant sets of scanners, allowing any three to override a fourth should it make an error. Should two scanners produce the same error the transport process would be aborted automatically and shut down by the transport controller system.

The pattern buffer is a large super conducting tokamak device, usually situated directly underneath the transporter unit itself. Once the subject has been dematerialized they are passed into the pattern buffer and held in suspension while the system compensates for relative motion between itself and the target location.

Pattern buffers can be shared by several different transport systems, although only one transporter can use a given buffer at a time. Should an emergency arise during transport a pattern can be held suspended in a transport buffer without being either sent or dematerialized; however, after a few minutes such a pattern will begin to degrade to the point at which the subject will be unrecoverable.

The biofilter is an image processing device which analyses the data from the molecular imaging scanner in order to locate any potentially damaging organisms which may have infected the subject. The biofilter is not generally a part of civilian transporter systems, though it is mandatory on all Starfleet transporters.

The emitter pad array is mounted on the exterior of the transport system itself – in the case of a spacecraft, on the hull of the ship. The array transmits the actual matter stream to or from the destination. Components of the emitter array include the phase transition matrix and primary energizing coils.

Some transporter systems also contain clusters of long range molecular imaging scanners within the emitter pad; this allows the system to lock onto targets at long range to beam them from remote locations without outside assistance. Most transporter systems do not include long range molecular scanners; such transporters can only beam to and from other similar systems.

Teleportation: From Star Trek To Tesla

Targeting scanners are a set of redundant sensors which are responsible for determining the exact location of the destination in relation to the transporter unit. Targeting scanners also determine the environmental conditions at the target site.

Although dedicated targeting scanners should ideally be a component of any transport process, in practice any sensor device of sufficient range and accuracy can provide the required information so long as it is compatible with the transporter controller information protocols. In addition, if transport is being conducted between systems with a fixed relative position – planetary transporter units, for example – targeting information can be disregarded.

The precise operation of a transporter naturally depends on the level of system specifications. Starfleet transporters are generally reckoned to be the most advanced in the Federation, since they are required to perform a wider range of tasks over much more variable conditions than civilian models. Typical operations for Starfleet transporters include the following.

Beam up involves using the emitter array as the primary energizing coil in order to beam a subject from a remote location which does not have a transporter system.

Site-to-site transport involves following the conventional beam up process until the subject is in the pattern buffer; the subject is then shunted to a second pattern buffer and on to another emitter array before being beamed out to a new location. This process essentially merges two transport processes in order to allow a subject to be beamed from one location to another without having to rematerialize on board ship first. This process is avoided if possible since it requires double the energy expenditure and system resources to accomplish each transport.

Hold in pattern buffer. As described, the pattern buffer can be used to hold a subject essentially is stasis. Normally these patterns will degrade after just a few minutes at most, though on one occasion a specially modified transporter held a subject intact for seventy five years.

Dispersal. Although transporter systems are designed to beam a subject to or from a destination intact, it is possible to override the safety systems on a standard Starfleet transporter and cause it to deliberately disperse the subject over a wide area. This is done by disengaging the annular confinement beam during rematerialization, depriving the subject of a proper reference matrix to form against. Such a measure may be used in order to neutralize a dangerous payload such as a bomb or other weapon; the measure is frequently complemented by materializing the subject in space.

Near warp transport is achieved by careful shifting of the ACB frequency. This can be an uncomfortable experience for those who go through it, and on occasion can even be dangerous. Warp transport can be achieved by the same method as near warp transport; this is only effective if the origin and destination are moving at the same warp speed.

Teleportation: From Star Trek To Tesla

Majel Barret-Roddenberry Reveals the Origins of the Transporters

Majel Barrett-Roddenberry, wife of the late *Star Trek* creator, Gene Roddenberry, has appeared in more episodes of the show than any other performer. She was first cast as "Number One" in the failed pilot of *Star Trek*, and later as Nurse Chapel, Dr. McCoy's long suffering assistant.

Finally, she returned in *Star Trek: The Next Generation* as Lwoxana Troi, the Betazoid whose antics often disrupted the crew of the Enterprise. But throughout the different shows, Barrett-Roddenberry has been the voice of the computer – announcing a self-destruct code or a greeting with equal indifference.

"No one had seen a computer, of course," she says. "They'd been in existence for 30 years but you couldn't use one. The only ones that were useable were monolithic monsters that they kept in the basement."

Star Trek had a large budget for its day but not enough to purchase actual high-tech equipment. Initially, the computer was to spew out information on scraps of paper, until producers realized that audiences would become bored with waiting. They substituted a voice, Majel's, in what now seems to be a prophetic look at today's computer developments.

"Most of our 'miracles' or 'discoveries' that the show is credited with were necessitated by the fact we had almost nothing to spend," she says. "The real reason we ended up beaming down by transporter was that Gene couldn't figure out a way to take the ship in and land it every week. Not only would that have been prohibitive because of money but it also would have slowed up the stories."

CHAPTER 4
GIFTS FROM THE WORLD OF SPIRITS

In the nineteenth centurty, Spiritualism was sweeping the planet as people huddled in darkened rooms attempting to contact the spirits of passed on loved ones. In these rooms, a strange phenomena was observed and recorded for posterity, the appearance and disappearances of all sorts of objects by mediums during seances.

Referred to as apports, these objects ranged from flowers, jewelry, to even live animals. The production of the apports was and is still one of the most prominent and effective aspects of the seances. Their behavior vary from flying through the air, to hitting the sitters in their faces, to landing on the table or in people's laps.

Many mediums claimed to be able to produce apports during seances, saying they were gifts from the spirits. However, more than a few mediums were discovered to be cheating as the objects, such as flowers, were found secreted within their clothing prior to the seance. There is, however, good evidence that apports do take place and many creditable witnesses have offered testimony to support this strange phenomenon.

The first recorded observation of apports appeared in the *Researches psychologique ou correspondencesur le magnetisme vital entre un Solitaire et M. Deleuze* (Paris, 1839). It was witnessed by a Dr. G. P. Billot during a seance on March 5, 1819. At the seance were three somnambules and a blind woman. One of the mediums said she saw a dove flying around the room. It was carrying something in its beak which it finally deposited before a person. When Billot examined the contents of the packet he found three pieces of paper with a small bone glued to each and beneath was written, "St. Maxime, St. Sabine and Many Martyrs."

Later Billot with the blind woman told of the experience to Dr. Deleuze who said he thought that animal magnetism was probably a better explanation than the intervention of spirits. There have, of course, been other explanations.

Apports have been recorded as arriving in various ways. During a seance in 1882, a pair of modest earrings given by a guide spirit to Marchioness Centurione Scotto came by means of a trumpet with a phosphorus band which appeared. The trumpet turned its large end up to fit against the ceiling and then there was heard a thumbing sound as the earrings dropped into the instrument.

In his work *Man's Survival After Death*, the Reverend C. L. Tweedale described an incident which involved his mother, wife and himself. His mother had suffered a cut on her head. They were all in the dinning room. His wife had just parted the older woman's hair to examine the wound. The minister suddenly looked and saw coming through the air with force from an opposite corner of the room above a window, behind his wife, was to a jar of ointment. The jar was one which his mother kept locked away in a chest. To him, the obvious indication was to apply the ointment to his mother's injury. Tweedale notes other similar incidents in his book.

There is the question from where do these apports come from. The question is legitimate when there is no fraudulent manipulation of the objects involved. Apports of flowers have been traced to a nearby garden. One incident involved Attorney Henry Olcott when attending a seance held by Helena Blavatsky, who was presented with a leaf of a rare plant on which he had previously put a mark.

Henry Olcott was born in New Jersey in 1832 and attended college in New York City, studying agricultural science. While still in his early 20's, he received international recognition for his work on a model farm and for founding a school for agriculture students. During this same time, he published three scientific works. He went on to become the farm editor for Horace Greeley's newspaper, the *New York Tribune*.

When the Civil War broke out, Olcott enlisted in the Union Army. He was appointed as a special investigator to root out corruption and fraud in military arsenals and shipyards. He was soon promoted to the rank of Colonel and after the war, was part of a three-person panel that investigated the assassination of President Lincoln. After the war, Olcott studied law and became a wealthy and successful lawyer.

An early example of seance involved teleportations investigated by Olcott involved the Eddy brothers of Chittenden, Vermont. According to newspaper and Spiritualist accounts of 1874, some very strange things were taking place in the home of William and Horatio Eddy, two middle-aged, illiterate brothers, and their sister, Mary.

The Eddys lived in a two-story building which was reported to be infested with a number of spirits. The events at the farm were said to be so powerful and so strange that people came from all over the world to witness them. Spiritualists began calling Chittenden the "Spirit Capital of the Universe."

After buying a copy of the Spiritualist newspaper *Banner of Light*, Olcott read with interest the reports from the Eddy farm. Although skeptical, he knew that if the stories were true, "this was the most important fact in modern physical science," he later wrote. A short time later, Colonel Olcott traveled to Vermont, accompanied by a newspaper artist named Alfred Kappes. Together, they planned to investigate the strange events at the Eddy farm and if the stories were a hoax, they would expose the Eddy brothers in the *Daily Graphic* newspaper.

If the Eddys were true mediums, Olcott would announce the validity of Spiritualism to the world. If they were fakes, he would expose them as nothing but charlatans. In either event, Olcott was determined to be fair in his judgments.

Olcott and Kappes traveled to the secluded town of Chittenden, located within the Green Mountains. The trip out to the farm was uneventful, but the first meeting with the Eddy brothers was anything but ordinary. The two distant and unfriendly farmers had New England accents so thick the New York attorney and writer could scarcely understand them.

Olcott would later learn that the brothers were descended from a long line of psychics. Mary Bradley, a distant relative, had been convicted of witchcraft at Salem in 1692. She had escaped the village with the help of friends. Their own grandmother had been blessed with the gift of "second sight" and often went into trances, speaking to entities that no one else could see, and producing apports of exotic fruits not commonly found in the area. Their mother, Julia, had been known for frightening her neighbors with predictions and visions although her husband, Zepaniah, condemned her powers as the work of the Devil. Julia quickly learned to hide her gifts from her husband.

However, the supernatural could not be hidden once the couple began having children. Strange knocks and bumps began shaking the house; disembodied voices were heard in empty rooms; and occasionally, the children even vanished from their cribs. They were likely to be discovered anywhere in the house and even outside as if teleported by unseen forces. As William and Horatio got older, their strange powers strengthened. On many occasions, Zepaniah would see the boys playing with unfamiliar children, who would vanish whenever he approached

The boys soon learned they were unable to attend school. The initial attempts were marked by inexplicable happenings and disturbances as invisible hands threw books, made paper, coal and coins appear and disappear, levitated desks and caused objects like rulers, inkwells and slates to fly about the room.

Zepaniah, tiring of his sons trances and antics, sold the brothers to a traveling showman, who for the next 14 years, took them all over America, Canada and Europe. As part of the performance, he would challenge audience members to try and awaken the boys from their trances, which of course they never did.

On several occasions, they were even stoned and shot at by audience members, convinced the brothers were fake. William Eddy bore a number of bullet scars as a result of their early show days. Only after their father died were the boys able to return home. They moved onto the family farm with their sister, Mary, and opened the house as a modest inn called the Green Tavern.

On Olcott's first day at the farm, he was witness to an outdoor seance. In the bright moonlight of a warm summer evening, a group of ten participants traveled down a path and into a deep ravine. They assembled in front of a natural cave, formed by two large stones which had collapsed atop one another, forming a large arch. Olcott later learned that it was called "Honto's Cave," in honor of the Native American spirit who often appeared there. Olcott suspiciously investigated the cave and saw that no exit could be found at the back of the rocks. He determined there was no way that anyone could slip in or out of the cave without being seen.

Teleportation: From Star Trek To Tesla

Horatio Eddy acted as the medium for the seance. He sat on a camp stool under the arch and then was draped in a makeshift "spirit cabinet" formed by shawls and wood cut from small saplings. As Horatio rested there, a gigantic man, dressed as a Native American, emerged from the darkness of the cave. While the medium addressed this spirit, someone cried out and pointed up toward the top of the cave. Standing there, silhouetted against the moon, was another gigantic Indian. To the right, another spectral female had materialized on a ledge. In all, ten such figures appeared during the seance.

The last, the spirit of a man named Alas Sprague, emerged from within Horatio's cabinet. He vanished at the same time the others did. Moments later, Horatio appeared from the cabinet and signaled that the seance was at an end. With him he brought gifts of flowers and arrow heads that he said were gifts from the spirits. After the seance was over, Olcott and Kappes carefully searched the cave and the surrounding area for footprints in the soft earth. They found no trace that anyone had been there.

Olcott found the seance to be convincing but was sure that he would be able to more easily detect fraud within the controlled setting of the Eddy house. He and Kappes thoroughly examined the large "circle" room, which was located on the second floor of the farmhouse. Using carpenters and engineers as consultants, the two men became convinced that the walls and floors were as solid as they seemed. Because of this, what Olcott witnessed during the nights that followed became even stranger.

Each seance was basically the same. On each night of the week, except for Sunday, guests and visitors would assemble on wooden benches in the seance room. A platform, which had been assembled there, was lit only by a kerosene lamp, recessed in a barrel. William Eddy, who acted as the primary medium, mounted the platform and entered a small cabinet. A few moments later, soft voices began to whisper in the distance. Often, it would be singing, accompanied by spectral music.

Musical instruments came to life and soared above the heads of the audience members; disembodied hands appeared, waving and touching the spectators; odd lights and unexplained noises appeared and filled the air. Then, the first spirit form emerged from the cabinet. They came one at a time, or in groups, numbering as many as 20 or 30 in an evening.

Some were completely visible and seemed solid. Others were transparent and ethereal. Regardless, they awed the frightened spectators. The spirits ranged in size from over six feet to very small. Most of the ghostly apparitions were elderly Yankees or Native Americans but many other races and nationalities also appeared in costume like Africans; Russians; Orientals; and more.

The apparitions not only appeared but they also performed, sang and chatted with the sitters. They also produced spirit articles like musical instruments, clothing and scarves. In all, nearly every type of supernatural phenomena was reported at the Eddy farmhouse.

Teleportation: From Star Trek To Tesla

These included rappings; moving physical objects; spirit paintings; automatic writing; prophecy; speaking in tongues; healings; unseen voices; levitation; remote visions; teleportation; and more. And of course, the full-bodied manifestations of which Olcott observed more than 400 during the weeks he visited the house. He concluded that a show like that which he had seen would have required an entire company of actors and several trunks of costumes.

The author's ten-week stay on the Eddy farm was surely a test of endurance. However, he was convinced of the fact that the two men could make contact with the dead. Not only did Colonel Olcott chronicle his visit in the newspaper, but he also wrote a massive book called *People from the Other World*. The book, over 500 pages long, is full of precise drawings of the apparitions, the grounds, the house and even detailed plans of its construction, proving that no hidden passages existed.

He also recorded over 400 different supernatural beings, registered dozens of objects that had been apported by the spirits, and collected hundreds of affidavits and scores of eyewitness testimony to the amazing events. He also reproduced dozens of statements from respected tradesmen and carpenters who had examined the house for trickery.

Eventually, the Eddy brothers and sister Mary, went their separate ways. Their bickering and feuding had driven them apart. Horatio moved out and took a house across the road, where he took up light gardening, occasional seances and doing magic tricks for local children. Mary moved to the nearby village of East Pittsford, where she became a full-time professional medium. William dropped out of public life altogether and became a bitter recluse on the family farm.

The first of the Eddys to die was Horatio on September 8, 1922. William lived for another 10 years. He never married and refused to ever participate in Spiritualism again. He died on October 25, 1932 at the age of 99. If either of the men had any secrets about the weird events at their home, they took the secrets with them to the grave.

The Strange Mediumship of Jack Webber

One of the most interesting mediums involved with apports and teleportations was the Welsh coal miner Jack Webber, whose death at the early age of thirty-three in March 1940 was a great loss to psychic research. Webber was a simple man, a coal miner who left school at the age of fourteen to go down the pits, where he toiled until a few years before his death.

Webber had no inkling of his own psychic powers, and treated all accounts of psychic phenomena with disbelief and scorn. It was only at the age of twenty-one when he first met his fiancee, who belonged to a staunch spiritualist family, that he began attending seances and in so doing learned that he himself possessed considerable psychic talents.

Teleportation: From Star Trek To Tesla

According to Professor John Frodsham, who has done extensive research on the subject, Webber discovered his own gifts through the simple process of table-rapping. It was not until two years after this that he was able to fall into a trance. A year or so after this Webber developed healing powers. Under trance thick oil would ooze from his hands. With this oil, which had the consistency of Vaseline, he would massage the patient who was more often than not cured of his affliction.

Often during the day, half-entranced, he would go out into the marshlands and open country near his home, gather certain herbs, return home and brew them into potions which he then administered to the sick. His healing powers were undoubted for he had many successful cures to his credit, but his ministrations exhausted him so much that they put a severe check on the physical phenomena which were then beginning to manifest themselves through him.

At this time he was still working down the mines, where he remained until 1936. The combination of intense physical labor, followed by fatiguing trance sessions at night reduced him to a state of near exhaustion. It may, in fact, have been these years of continuous, over-exertion that brought about his sudden and premature death.

In any case, he was soon compelled to give up both his work as a healer and his toil in a Welsh coal mine in order to devote himself entirely to physical mediumship. Curiously enough, Webber was for many years afraid of the physical phenomena that built up around him.

At night when he went to bed, loud rappings would be heard, the bed clothes would be ripped off him and objects would fly wildly around the bedroom while voices spoke and muttered around him in the darkness. Only when he fell into trance and still more powerful phenomena appeared did Webber lose his fear of these manifestations. It was not until shortly before his death that he was able to accept the physical phenomena without fear.

Webber's mediumship aroused keen public interest. In the fourteen months from November 1938 to the end of December 1939 he gave over 200 demonstrations of his powers at seances attended by over 4,000 people in conditions ranging from public exhibitions to demonstrations held in his own home.

Since Webber never used a cabinet, numerous photographers were able to obtain excellent infra-red and flash photographs of ectoplasm. Strings and sheets of this substance, often many yards long, are seen in these photographs to be emerging from the medium's mouth and fanning out on the floor in front of him where they are often held up for inspection by the sitters.

Webber was invariably tied to a chair with a rope some fifteen yards long. When the tying-up had been completed the ends of the rope were sewn up, sealed with wax-and then impressed with a seal provided by a sitter. A piece of cotton was then tied at the base of Webber's thumb, a piece of paper threaded on through a needle hole and the other end of the cotton tied to the base of his opposite thumb. This made it quite impossible for him to move his hands more than a few inches.

Such precautions did not stop the rope from being removed by unseen hands during the seance. On one occasion the rope was removed from the chair, taken right across the circle, and tied around the chair of a sitter opposite, being wound under the seat and around the chair legs.

When the seance was over the rope had to be cut through to free the unfortunate sitter. On several other occasions, a few seconds after Webber had been roped in the chair the lights were switched on only to show the medium standing on the far side of the circle with the ropes resting on the chair precisely as tied. This was astonishing since it normally took close on five minutes for two people to tie Webber securely to his chair and much longer to untie him.

Even more surprising was the fact that shortly after Webber had been seen standing on the other side of the circle he would begin to spin around rapidly, the lights would be put off and five seconds later he would be found back in his chair roped precisely as before with the cotton and the sealing wax unbroken.

It should be noted that such performances were witnessed by literally hundreds of reliable witnesses, all of whom were prepared to swear to the supernatural character of the phenomena they had seen: In June 1939, for example, three highly-placed representatives of the BBC were present at a seance in which they themselves tied the medium into his chair and fastened his coat to him with cotton, yet during the seance, even while the medium's hands were being held by two of these gentlemen, Webber's jacket was taken off his back without the cotton being broken. A few seconds later the coat was returned to Webber with the cotton intact and knotted around the coat buttons as it was in the beginning.

Webber is well known for his astonishing production of ectoplasm which flowed from him in great quantities, always under strict control conditions. In his book devoted to Webber the well-known British healer, Harry Edwards, points out that Webber produced two types of ectoplasm – namely ectoplasmic arms and ectoplasmic rods. These arms were used to apport objects – for Webber was famous for his apports – as well as to construct voice boxes which either emanated from the medium or were attached to the trumpets.

The arms were soft and flexible though coarse in texture. They were equipped with tentacles at the end which could be used for moving objects. At times the ends of these arms were self-illuminated by a blue ring of light with a dark center. These lights, which strongly resemble those produced by earlier mediums, first appeared near Webber's solar plexus and then moved out to his sides and above his head. They were at all times responsive to the commands of the Guide.

The ectoplasmic rods were generally invisible and could not be photographed. Nevertheless, these rods were sometimes seen by sitters when a little daylight was allowed to filter through the window. Edwards described them as strong, thick, straight structures, three to six inches in circumference, which attached themselves to any levitated or teleported object.

In February 1940, during a routine sitting, Webber's jacket was dematerialized in such a way that while the coat sleeves remained on his arms under the ropes with the shoulders and lapels of the coat in their proper position, the back of the coat was draped across the front of the body.

Webber himself believed that the back part of the jacket had been dematerialized and brought through the body like an apport and then rematerialized across his chest. This surprising feat was accomplished in the space of a few seconds while the medium was in deep trance and roped firmly to his chair.

The most notable of Webber's apports was at a public sitting in Paddington on November 8, 1938, at which a brass bird was apported from a neighboring room. Immediately before this occurred, Webber's control had told the chairman of the sitting, Harry Edwards, that the entities intended to bring a brass ornament in the form of a crane into the room through the medium's body and would permit it to be photographed during the process of teleportation.

Photographs show the bird in question, an ornament three inches high and weighing two ounces, emerging from the medium's body wrapped in ectoplasm.

At other Webber seances several similar articles were also apported, among them a small stone statuette of Buddha, a mosaic ornament in the form of a brooch and an Egyptian seal depicting Osiris. It is interesting that all these objects, including the crane, share a common Eastern and occult significance. Furthermore, all of them with the exception of the crane, were of unknown origin. They simply materialized from Webber's body.

Prior to their arrival he had sensed a tightness in his abdomen which indicated to him that the process of teleportation was about to begin. He had therefore asked to be searched in front of the sitters just before he was roped to the chair. It is obvious from the very size of the objects concerned; that none of these apports could possibly have been hidden about the medium's person and that they must all have arrived in much the same manner as the crane. These were not by any means the only apports that Webber produced during his seances but they are certainly the most spectacular.

Webber's feats of teleportation put him in the same class as the greatest mediums. Crookes himself, working with D.D. Home and others, witnessed the arrival of apports on no less than twelve occasions.

Stainton Moses, on August 28, 1872, heard a small hand-bell move from a neighboring room ringing loudly, pass through a closed door and finally, after completing a circuit of the room, materialize on the table close to his elbow. Enrico Morselli (1852-1929) Professor of Psychiatry at Genoa University, witnessed during the course of thirty sittings with Eusapia Paladino, the sudden appearance of objects that had obviously not originated from inside the room, or for that matter, anywhere close by. Items such as flowers, branches, leaves, nails, coins and stones would make their way through solid walls and closed doors.

Dr Julien Ochorowitz (1850-1918) when working with Stanislawa T. frequently observed the disappearance and reappearance of objects in full light. Madam d'Esperance produced spectacular apports of flowers.

On August 4, 1880, she caused a plant twenty-two inches in height with twenty-nine leaves, all of them smooth and glossy, to appear in a water carafe which it filled so completely that it could not be removed. In the photographs which were taken of this plant almost immediately after its appearance, it can be seen that the roots were wound around the inner surface of the glass as though they had germinated on the spot and never been disturbed.

On June 28, 1890, the same medium apporting a golden lily seven feet in height bearing eleven large blossoms. After the plant had been photographed by Professor Boutleroff it vanished as mysteriously as it had come leaving behind only a couple of fallen blossoms.

Medium Carlos Mirabelli (1889-1951), a Brazilian of Italian parentage, has been attributed with a good number of apports and teleportations during his long career. Carlos was once reported to have dematerialized in broad daylight before a crowd of people, and reappeared ninety kilometers away. Witnesses report that Mirabelli suddenly went into a trance and in full view of several hundred people, slowly grew transparent and faded from view. When he reappeared, he told his friends that the spirits had taken him away.

There were also instances of Carlos dematerializing from the sealed seance room to another room, and the seals on his bonds being found untouched. When he disappeared, some of the sitters remained in the seance room while others went to search for him. He was soon discovered in a side room lying in an easy chair, still in a complete trance and singing to himself.

A more recent report of apports during seances comes from John G. Sutton, who in January of 1999 attended a public demonstration of mediumship at The Vogue Theater on Hollywood Boulevard, Hollywood, California. While in the theater there appeared, from out of thin air, a number of American coins, mainly dimes and one cent pennies.

These coins dropped onto the clean red carpet leading down through the auditorium towards the stage. The theater management explained that this was a regular phenomenon and no matter how often they cleaned the carpets coins would be found upon it.

Sutton writes that he was walking through the theater towards the stage with all the interior lights on, and the carpet was definitely clear. Less than two minutes later on walking back towards the box office he saw a one cent coin in the middle of the carpet. This he picked up and placed in his pocket. It had not been there when he walked down, no one had followed him, there was no balcony in the theater, so how the coin arrived there remains a mystery. Perhaps it was teleported from another dimension, or from someone's change jar.

41

CHAPTER 5
PENNIES FROM HEAVEN

Possibly the strangest events in the annals of extraordinary phenomena are the mysterious rains of objects not normally expected to be found falling from the clouds. Numerous newspaper reports and books have been written describing strange things seen coming down from the sky.

Such diverse items as fish, frogs, monkeys, pieces of raw meat and blood, chains, coins, common gravel and more have all been described to have been seen falling from the rainy skies. Such strange falls are often attributed to tornados sucking up things into the heavens and then dropping their payload miles from their point of origin. However, this explanation is usually unsatisfactory as most unexplained falls are never accompanied by other storm debris.

Sky falls have been recorded since the beginning of recorded history. The Bible's the Book of Numbers states: "When the dew fell upon the camp at night, the manna fell upon it." We are told that, "the manna was as coriander seed, and the color thereof as the color of bdelium." (A gum from Commiphora trees.)

In fact, the original Hebrew word, for which bdelium is given as a translation, is b'dolakh, the meaning of which is unknown. Modern speculation has suggested that manna was honeydew, a sweet excretion of plant lice (aphidae) and scale insects (coccidae). According to the Old Testament, the Israelites baked it into cakes, which tasted of fresh oil.

The ancient Greek explanation for the most common creature falls involving fish and frogs was echoed by Pliny the Elder (23-79AD) who said that frog and fish seed in the soil was brought to life by rain. The scientific revolution in the 17th Century divided sky falls into the acceptable – like rain, snow and hail, which could be explained in terms of physics – and all the rest, which were dismissed as superstition, hallucination and tall stories. The scientist and animal expert, Ivan T. Sanderson, coined a useful acronym for "Things that are alleged to fall from the skies," he called them **Fafrotskies.**

Pioneer writer of the unexplained, Charles Fort, delighted in writing about such unusual atmospheric events in his groundbreaking books of the weird world we live in. In his *The Book of the Damned* and *LO!*, Fort gleaned stories from newspapers from around the world recounting the vast amount of things seen falling from the sky.

A good example of these types of fall took place in 1877. Thousands of live snakes, from one foot to 18 inches in length, rained down on the southern town of Memphis, Tennessee. *Scientific American* (36:86) tried to give a rational explanation: "They probably were carried aloft by a hurricane and wafted through the atmosphere for a long distance."

Teleportation: From Star Trek To Tesla

In Timb's *Year Book*, 1842-275, it is said that, at Derby, the fishes had fallen in enormous numbers; from half an inch to two inches long, and some considerably larger. In the *Athenum*, 1841-542, copied from the *Sheffield Patriot*, it is said that one of the fishes weighed three ounces. In several accounts, it is said that, with the fishes, fell many small frogs and pieces of "half-melted ice."

We are told that the frogs and the fishes had been raised from some other part of the earth's surface, in a whirlwind; no whirlwind specified; nothing said as to what part of the earth's surface comes ice, in the month of July. What interests us is that the ice is described as "half-melted." In the *London Times*, July 15, 1841, it is said that the [175/176] fishes were sticklebacks; that they had fallen with ice and small frogs, many of which had survived the fall. We note that, at Dunfermline, three months later (Oct. 7, 1841) fell many fishes, several inches in length, in a thunderstorm. (*London Times*, Oct. 12, 1841.)

London Times, Jan. 13, 1843 – according to the Courrier de l'Isére, two little girls, last of December, 1842, were picking leaves from the ground, near Clavaux (Livet), France, when they saw stones falling around them. The stones fell with uncanny slowness. The children ran to their homes, and told of the phenomenon, and returned with their parents. Again stones fell, and with the same uncanny slowness. It is said that relatively to these falls the children were attractive agents. There was another phenomenon, an upward current, into which the children were dragged, as if into a vortex. We might have had data of mysterious disappearances, but the parents, who were unaffected by the current, pulled them back.

In the *New York Sun*, June 22, 1884. June 16th – a farm near Trenton, N.J. – two young men, George and Albert Sanford were hoeing in a field when stones began falling all around them. There was no building anywhere near, and there was not even a fence behind which anybody could hide.

The next day stones fell again. The young men dropped their hoes and ran to Trenton, where they told of their experiences. They returned with forty or fifty amateur detectives, who spread out and tried to observe something, or more philosophically sat down and arrived at conclusions without observing anything. Crowds came to the cornfield. In the presence of crowds, stones continued to fall from a point overhead. Nothing more was found out.

In the *Journal of the Society for Psychical Research*, 12-260, is published a letter from Mr. W.G. Grottendieck, telling that, about one o'clock, one morning in September, 1903, at Dortrecht, Sumatra, he was awakened by hearing something fall on the floor of his room. Sounds of falling objects went on. He found that little, black stones were falling with uncanny slowness from the ceiling, which was made of large, overlapping, dried leaves.

Mr. Grottendieck writes that these stones were appearing near the inside of the roof, not puncturing the material, as if they were passing through this material. He tried to catch them at the appearing point, but, though they moved with extraordinary slowness, they evaded him. There was a coolie boy, asleep in the house, at the time. "The boy certainly did not do it, because at the time that I bent over him, while he was sleeping on the floor, there fell a couple of stones." There was no police station handy, and this story was not finished off with a neat and fashionable cut.

Fort writes in *LO!* that: "If there is teleportation, it is in two orders, or fields: electric and non-electric – or phenomena that occur during thunderstorms, and phenomena that occur under a cloudless sky, and in houses. In the hosts of stories that I have gathered – but with which I have not swamped this book – of showers of living things, the rarest of all statements is of injury to the falling creatures. Then, from impressions that have arisen from other data, we think that the creatures may not have fallen all the way from the sky, but may have fallen from appearing points not high above the ground – or may have fallen a considerable distance under a counter-gravitational influence.

"I think that there may be a counter-gravitational influence upon transported objects, because of the many agreeing accounts – more than I have told of – of slow-falling stones, by persons who had probably never heard of other stories of slow-falling stones, and because I have come upon records of similar magic, or witchcraft, in what will be accepted as sane and sober meteorological observations."

Co-author of this book, Tim Swartz, had his own experience with a mysterious stone-thrower while investigating a poltergeist case near Springfield, Ohio in 1983. The reportedly haunted house was located between Dayton and Springfield and had been experiencing strange incidents for about two months. The family, two adults and two teenagers, had been host to almost an unlimited amount of weird activity, ranging from loud knocks on the walls, the moving of furniture, out of place puddles of water, and the appearances of stones that seemed to fly through the walls and ceiling.

"One evening while I was visiting the family, stones began to fall around the living room. Each stone appeared to slowly float down from the ceiling, but would then hit the floor with a loud crack. In less than ten minutes over fifty rocks were collected."

Swartz remembered a similar incident reported by the late Ivan Sanderson where he marked the rocks with a white X and threw them back into the night. The stones all returned.

"Seeing if this poltergeist could emulate Sanderson's," Swartz stated, "I marked ten rocks with a large X and threw them out the back door into the cornfield behind the house. Sure enough, in less than five minutes, all ten rocks were seen to float down from the ceiling. This experiment was repeated several times, when the poltergeist, obviously tiring from the game, stopped returning the rocks."

It should be noted that all paranormal activity stopped about a week after this incident. Possibly the poltergeist used up all of its available energies playing fetch with rocks.

Charles Fort speculated that there could be a sort of Sargasso Sea floating high above our heads that would occasionally drop debris down on us mortals below. Of course when Fort reported on such things as strange disappearances, he wondered whether something that mysteriously appears somewhere had not mysteriously disappeared somewhere else.

The *Annals of Electricity*, 6-499, reported that in Liverpool, May 11th, 1842 – "not a breath of air when suddenly clothes on lines on a common shot upward. They moved away slowly. Smoke from chimneys indicated that above ground there was a southward wind, but the clothes moved away northward."

There was another instance, a few weeks later as reported in the *London Times*, July 5th, 1842 – a bright, clear day, at Cupar, Scotland, when a women hanging out clothes on a common heard a sharp detonation, and the clothes on lines shot upward. Some fell to the ground, but others went on and vanished. Though this was a powerful force, nothing but the clothes it seized was affected.

In June, 1919, at Islip, Northampton, England, an occurrence like the occurrences at Liverpool and Cupar. A loud detonation was heard and a basketful of clothes shooting into the air. Then the clothes came down. Fort says that "There may be ineffective teleportative seizures."

London Daily Mail, May 6, 1910 – phenomenon near Cantillana, Spain. From ten o'clock in the morning until noon, May 4th, stones shot up from a spot in the ground. Loud detonations were heard. "Traces of an extinct volcano are visible at the spot, and it is believed that a new crater is being formed." But there is no record of volcanic activity in Spain, at this time – nor at any other time.

Atmospheric Wildlife

Everyone knows that fish, along with their watery neighbors, live in rivers, ponds, lakes and oceans. As far as it is known, they do not live in the sky. So why are there so many reports of fish, frogs, eels and other slippery characters, falling from the sky like so many raindrops?

Fort said that he received letters about strange appearances of living things in tanks of rain water that seemed inaccessible except to falls from the sky. Mr. Edward Foster, of Montego Bay, Jamaica, B.W.I., wrote to Fort to tell of crayfishes that were found in a cistern of rain water at Port Antonio, Jamaica.

In the *London Daily Mail*, Oct. 6, 1921, Major Harding Cox, of Newick, Sussex, tells of an appearance of fishes that is more mysterious. A pond near his house had been drained, and the mud had been scraped out. It was dry from July to November, when it was refilled. In the following May, this pond teemed with tench. One day, 37 of them were caught. Major Cox reviewed all conventional explanations, but still he was mystified. What makes this story interesting was his statement that tench had never been caught in this pond.

Teleportation: From Star Trek To Tesla

Fort wrote that eels are mysterious beings. It may be that what are called their "breeding habits" are teleportations. According to what is supposed to be known of eels, appearances of eels anywhere can not be attributed to transportations of spawn. In the *New York Times*, Nov. 30, 1930, a correspondent tells of mysterious appearances of eels in old moats and in mountain tarns, which had no connection with rivers. Eels can travel over land, but just how they rate as mountain climbers it is not known.

In the ***American Journal of Science***, 16-41, a correspondent tells of a ditch that had been dug on his farm, near Cambridge, Maryland. It was in ground that was a mile from any body of water. The work was interrupted by rain, which fell for more than a week. Then, in the rain water that filled the ditch, were found hundreds of perch, of two species.

The fishes could not have developed from spawn, in so short a time: they were from four to seven inches long. Nothing is said of dead fishes lying upon the ground, at sides of the ditch: hundreds of perch arrived from somewhere, exactly in this narrow streak of water. There could have been nothing so scattering as a "shower."

Fort speculates that it looks as if to a new body of water, vibrating perhaps with the needs of vacancy, there was response somewhere else, and that, with accuracy, hundreds of fishes were teleported. If somebody should have faith, and dig a ditch and wait for fish, and get no fish, and then say that we're just like all other theorists, we explain that, with life now well-established upon this earth, we regard many teleportations as mere atavisms, of no functional value.

One Froggy Evening

Along with fish, one of the more common animals seen to fall from the sky is frogs. While frog and toad rainstorms are extremely rare, they can actually happen. One of these rare events is described in the August 3, 1883 edition of the ***Decatur Daily Republican***.

"Cairo, Illinois, August 3 – Early yesterday morning the decks of the steamers **Success** and **Elliot**, moored at the Mississippi levee, were observed to be literally covered with small green frogs about an inch in length, which came down with a drenching rain which prevailed during the night. Spars, lines, trees and fences were literally alive with the slimy things, while the lights from the watchmen's lanterns were obscured by the singular visitation. The phenomenon, while not entirely unknown, has never been explained, and is causing considerable comment."

Another report about one of these bizarre events appeared in *Scientific American* magazine, dated July 12, 1873. It states, "A shower of frogs, which darkened the air and covered the ground for a long distance, is the reported result of a recent rainstorm at Kansas City, Missouri."

Teleportation: From Star Trek To Tesla

These reports are not just talking about a few frogs falling from the sky; they appear to be describing a downpour. What's more, frogs are not the only creatures which have been reported falling from the sky over the years. A 1930 issue of the magazine *Nature* reads, "During a severe hailstorm in Vicksburg, a gopher turtle, 6 inches by 8 inches, and entirely encased in ice, fell with the hail."

An interesting report by one witness appeared in a 1939 issue of an English journal called *Meteorological Magazine*. It recounts, "Mr. E. Ettles, superintendent of the municipal swimming pool, stated that about 4:30 p.m. he was caught in a heavy shower of rain and, while hurrying to shelter, heard behind him a sound as of the falling of lumps of mud. Turning, he was amazed to see hundreds of tiny frogs falling on the concrete path around the bath. Later, many more were found to have fallen on the grass nearby."

What could cause this strange shower of creatures? Is there some natural phenomenon taking place where schools of aquatic animals are being teleported from one location to another? Sort of a cross-dimensional migration. In the October 31, 1942 *Buffalo Evening News*, the British Information Service gives the following explanation of a shower of frogs. "The entire contents of small ponds are sucked up by a certain type of wind – in the same way that sand is sucked up in a sandstorm, which carries the moisture and the frogs a little distance before they fall to the earth again."

Dr. William Hayden Smith, Professor of Earth and Planetary Science at Washington University in St. Louis, believes these events are peculiarities of weather. "These frog and toad rainfalls are most likely caused by tornadoes or violent thunderstorms," he explained.

"The heavy winds will pass over ponds and creeks and pick up small creatures from the water. They will pull these animals up to high levels in the atmosphere and carry them as the tornado covers hundreds of miles of land. Then, they will drop them to the ground."

When asked about these falling creatures, University of Missouri-Columbia climatologist, Patrick Guinan, compared these occurrences to the July 4, 1995 tornado that swept through Moberly. People north of Keokuk, Iowa found many unopened soda cans that the tornado had lifted from the Double Cola Bottling Plant in Moberly and dropped about 150 miles north near Keokuk. Guinan says these cans falling from the sky are just like raining frogs.

Of course the tornado explanation does little to explain the selective nature of skyfalls. Usually there are no reports of other debris falling with the fish or frogs. You would think that if a tornado picked up the contents of a pond – mud, weeds, rocks and wildlife – they would all fall together. But it just doesn't happen that way.

The citizens of Naphlion, a city in southern Greece, were surprised one morning in May, 1981, when they awoke to find small green frogs falling from the sky. Weighing just a few ounces each, the frogs landed in trees and plopped into the streets. The Greek Meteorological Institute surmised they were picked up by a strong wind. It must have been a very strong wind. The species of frog was native only to North Africa. This story is common to others with multiple animal falls. Usually the creatures and not native to the area of the fall.

The *Fortean Times* published a first hand account of a frog fall in its first issue, back in 1973. Mrs S. Mowday went to see a Royal Navy display on the Meadow Platt in Sutton Park, near Birmingham, on June 12th, 1954.

"I attended the display with my young son and daughter. It was a Saturday and there were frequent heavy showers...We tried to shelter from a shower under the trees...when we were bombarded by tiny frogs, which seemed to come down with the rain. There were literally thousands of them. They descended on our umbrellas, on us and we were afraid to walk for fear of treading on them." A British frog fall also occured in October 1987, when large numbers of rose colored albino frogs fell on several occasions around Stroud, Cirencester and Cheltenham.

In 1995, *Fortean Times Online* reported that Nellie Straw of Sheffield, England, was driving through Scotland on holiday with her family when they encountered a severe storm. Along with the heavy rain, however, hundreds of frogs suddenly pelted her car. The frogs were all tiny little things, less than an inch across, and so colorless they almost seemed transparent.

All sorts of weird things have been seen to fall from the heavens, not just living animals. A three-inch metal rod, one inch in diameter, came within inches of killing production-line worker John MacGregor when it crashed through the roof of the Symphony Furniture factory in Leeds in early March 1995. Said Mr MacGregor: "It made a hell of a din as it came through, then another clatter when it hit the floor. It gave me quite a shock. It is certainly big enough to have killed me."

In early March 1983, Rita Gibson of Topsham, near Exeter, found a scattering of strange pink beans in her back garden. They were larger than rice grains and smaller than orange pips. "They could not have been thrown," she said, "because our house is surrounded by three walls around a courtyard." Though they were out of season, they were quite fresh.

In the Italian province of Macrata. Shortly before sunset, the sky filled with a large number of blood-colored clouds. Half an hour later, a cyclone storm burst, and the air was filled with millions of small seeds, which covered the ground to a depth of half an inch. They belonged to the genus Circis, commonly called the Judas Tree, found only in central Africa or the Antilles. A great number were in the first stage of germination.

During the night of 8th-9th of November, 1984, East Crescent, Accrington, in Lancashire, was bombarded with apples, best quality Bramleys and Coxes. Derek and Adrienne Haythornwhite found at least 300 on their back lawn, on the path and in the hedges, and more were discovered in nearby gardens. The couple were woken up in the night by thunderous noises on the roof. Adrienne said: "They kept on falling for an hour or longer." Maybe nature has a way to spread its flora and fauna by ignoring the laws of time and space.

Above: Several rocks that were marked by co-author Tim Swartz and returned by an unseen force during a poltergeist investigation in 1983.

Below: Typical poltergeist activity includes psychokinetic events such as the mysterious levitation and even teleportation of household objects.

CHAPTER 6
RECOLLECTIONS OF AREA 51 - BY COMMANDER X

By now everyone has heard of Area 51 – that top secret air base a hundred plus miles from Las Vegas – where all of those crashed UFOs were supposedly taken, studied and copied. Oh it wasn't a very good secret, the Soviets knew about it. Aviation experts knew about it. Only the general public were kept in the dark about the secrets located beyond the mountains of the Nevada desert.

Area 51, also known as Groom Lake and Dreamland, is but the tip of an iceberg, whose "main body" lies hidden in Nevada's jumbled mountain ranges. Beneath the Air force's top secret base experiments are taking place with aircraft which would appear to be of exotic technology and possibly extraterrestrial.

Only nine miles south of the Groom Lake facility, which has the longest runway in the world is S4, another top secret military facility adjacent to the Papoose Dry Lake. It is said to house nine captured alien craft. Whether they were created elsewhere on a planet far away or built with U.S. and German technology remains a mystery. What we do know for certain is that strange things have been seen flying in the dark skies above the mysterious base.

All of this comes to no surprise for me, after all, I have been to Area 51 numerous times in my long career in military intelligence. What shocks some who are not familiar with the daily activities at Area 51 is the amount of people in both the military and civilian sectors who work there. Many are flown in every day by charter jets operating out of Las Vegas.

These people go to and from their jobs everyday with little knowledge of the history and controversy surrounding their place of employment. It was here, in 1955, that the U-2 spyplane first took wing. In the years that followed, its successors, the A-12 and SR-71 and later the stealthy F-117A fighter and B-2 bomber, flew across the same clear Nevada sky.

In 1953, when Major John Seaberg, an aerospace development engineer at Wright-Patterson AFB, came up with an idea for a very fast jet that could overfly the USSR at 70,000 feet, the customary test site for such aircraft at the time was Edwards AFB. However, for a project as top secret as this one, security at Edwards was thought to be inadequate. Johnson told his top test-pilot, Tony LeVier, to go and find a secure site somewhere in the Southwest from which to test fly the plane, which was being called Aquatone by the CIA, and Angel by Lockheed.

LeVier and Dorsey Kammerier, another Lockheed employee, took off in a Beech Bonanza looking for dry lakes, which provided a ready surface for landing. They checked a dozen or so before they came to Groom Lake, Nevada. It was northeast of Las Vegas and adjacent to the Atomic Energy Commission Proving Grounds (later renamed the Nevada Test Site), and

Teleportation: From Star Trek To Tesla

had been used for target practice during World War II. The area was unpopulated, which was good because it had been sprayed regularly with radioactive fallout from the atomic tests at the proving grounds. There was a lead mine operated by the Sheahan family in the mountains near the dry lake, but work there was sporadic due to the nuclear tests.

The location seems like a bad choice for an air base because of the fallout. In fact, the cancer rates of those who worked at the base during its early days were staggering. But the fallout was part of the attraction. The security of the Proving Grounds was tight, and the fallout helped keep people away.

The area controlled by the AEC was expanded to include Groom Lake, and by July, 1955 the CIA had its secret base. A fake construction firm, CLJ, was invented to oversee the construction, which was mostly done by subcontractors. Hangars, a mile-long runway, a concrete ramp, a control tower, a mess hall, and other amenities were constructed. In official records, the base was referred to as Watertown Strip, but it was called Paradise Ranch or just the Ranch by the pilots and ground crews.

The first prototype of the top secret Aquatone was called: Article 341, and it was flown out to Groom on a transport plane from Lockheed's Burbank, California facility on 7/24/55. It arrived disassembled and wrapped in cloth. The plane had only two landing wheels; it was like landing a bicycle.

On its maiden flight, LeVier wanted to touch the rear wheel first, but Johnson insisted the best way was to touch the nose wheel down first. After two failed attempts at doing it Johnson's way, Levier landed it perfectly, rear wheel first, just before a rainstorm flooded the dry lake to a depth of two inches.

Legend has it that when LeVier got out of the cockpit, he saluted Johnson with a "one-fingered" salute for nearly getting him killed with his insistence on a nose-first landing. Johnson is supposed to have returned the "one-fingered" salute and yelled, "You, too!" The story was widely told among the pilots, and the plane itself became known as the "You, too!" or U-2. Unfortunately, I cannot either confirm or deny the validity of this story. But it does make a good tale to tell around a few bottles of beer after a hard day's work.

The thing to remember about all this is that the Ranch was not a U.S. Air Force Base. Lockheed and the CIA were in charge here, not the military. To be sure, U-2 pilots were recruited from among F-84 pilots with top-secret clearances from SAC bases. If they took the job, however, they resigned from the Air Force and became CIA employees.

The production U-2's were flown in from a small Lockheed factory at Oildale, California on board C-124 transport planes. The C-124 pilots weren't even told their destinations, but were told to fly (at night) to a certain point on the California-Nevada border, and from there they were directed by radio into their unknown landing site. Everyone turned in their regular I.D.s on arrival at Groom and used aliases while on duty. This procedure was still in operation the times I visited the base, and I am sure it is still used today.

Teleportation: From Star Trek To Tesla

Area 51 has been used for test flights of captured foreign aircraft, such as the MIG-21, and for testing the Stealth B-2 bomber, but what is being tested there now? There are rumors of craft like the TR-3A Black Manta, the Pumpkin Seed, and the now famous Aurora. Some kind of aircraft (maybe the Aurora, maybe not) has been seen that leaves a trail that looks like "doughnuts on a rope."

With my career in military intelligence I have actually had the opportunity to "cross over" from time to time to work with the NSA, CIA and a couple of other groups who are so top secret that I never knew their real names. This does occur with agents in my former line of work. And more often then you would think. Our skills are valuable and are needed from time to time by other groups. After all, we were all working for the same team.

Because of my "specialities," my services were requested almost a dozen times out at Groom Lake during the sixties and seventies. Some of the projects that I worked on involved exotic aircraft that looked just like the stereotypical flying saucer. Needless to say, the first time I ever laid eyes on these amazing flying machines, I was shaken to the very core of my soul.

What must be understood is that I was never given any explanation on what was really going on at Area 51. I had a job to do and that was it. I was given no information that was not related to my particular job. If I had any questions that did not fit in with my job description, they were ignored. If I pushed a little too hard, I was taken away and debriefed on why I needed to ask my question. It was made perfectly clear that the wrong questions could lead to dismissal, a court martial, or even worse. I didn't want to find out what "even worse" meant.

As with any military or intelligence operation, the rumor mill was all too willing to supply gossip and innuendo on the true nature of the various projects going on around the base. Looking back in retrospect, it is amazing how many of these stories made it out of these supposedly secured bases and into the eager hands of civilian investigators. Not to say that these tales were true or not. I am just saying that the same strange tales that I heard on base in the sixties and seventies, later resurfaced in the eighties and nineties in magazine articles and books dealing with conspiracies and UFOs.

For example, it was impossible not to consider that the flying saucer shaped aircraft that I saw (and even briefly flew), were not extraterrestrial in origin. And the rumor mill confirmed such speculation. The craft were allegedly alien spacecraft that had crashed or were obtained by some unknown means. However, can I say for sure that these rumors were true? I cannot. With the exception of the craft themselves, I saw no substantial proof that they were extraterrestrial spacecraft.

Knowing what I do about the fine art of disinformation, the whole "UFOs are alien-driven spacecraft" could just be a huge hoax to disguise the fact that we are flying some sophisticated aircraft using technology that is years ahead of what is known by civilians. Or it could be true. This is what makes disinformation work so well. A little truth here, a little lie there. Keep them confused with wild tales, fake informants, a couple of authentic secret documents, surrounded by dozens of fakes, all work to hide the truth to all but a select few.

There were only a few times at Area 51 that I had any contact with the saucer aircraft. I had been trained a few years earlier for possible remote-viewing operations. I was fairly-good at remote-viewing, but obviously my trainers felt I had more potential because I was asked to come to Groom Lake (Area 51), and to try and fly one of their secret aircraft.

I tried to tell them that they were mistaken, that I was not a pilot, but I was told to come out anyway. I arrived early in the morning, flown in by jet from McCarran airport in Las Vegas. I was immediately loaded onto a bus with the windows blacked out and sent to an area about 15 miles from Area 51 and hidden in the mountains.

When we finally stopped and were let off, I found myself looking at the base of what I later found out was Papoose mountain. On close inspection I could see a series of large doors (which I later found out were hanger doors), that had been built into the base of the mountain and camouflaged in such a way as to blend right in with the color and texture of the soil and rocks. I later found out that I was at a secret location called S-4.

There were nine hangers with the entrance hidden in a natural indentation. Some reports mention extensive security at this location. However, for me, there were no security checks and I saw no armed guards or apparent military personnel. Inside these camouflaged hangers was the stuff that dreams are made of. To my eyes it was a classic UFO, a flying saucer.

Bob Lazar has said that he was taken to an area that sounds very much like the place I was in. The exception being that Lazar claimed that he saw several different kinds of flying saucers hidden away under the hills. I only saw one. But one was enough.

The flying saucer before me was shaped like a straw hat. It had a thin round base upon which sat a squat cylinder. It looked exactly like the Rex Heflin UFO photographs taken in the 1960's. I have often wondered if this wasn't the very same UFO photographed by Heflin along that lonely stretch of highway in california.

Its color was a dull silver, almost pewter. It looked dark despite the banks of lights that shone down from the girders above. It looked as if it were absorbing the light instead of reflecting it. I was later told that when it flight, it would change colors depending on its speed and proximity to the ground.

My job was simple. I was to get aboard, and with the power of my mind, I was to attempt to fly this unusual craft. That was all I was told. I was not allowed to ask any further questions. I was to simply shut up and to try and fly this thing.

If this craft were made by aliens from another planet, there was no way of telling from the exterior of the saucer. It was entirely bare except for a single chair that was bolted to the center of the floor. The room was cylinder shaped about ten feet high and twenty feet wide. It was well lit, except there was no apparent light source.

Teleportation: From Star Trek To Tesla

The chair was built for a normal sized human and had numerous cables connected to it that disappeared into the silvery gray floor. At various points on the chair, the seat, the armrest, the headrest, were small metal plates that were wired into the larger cables.

The craft was then pushed through the hanger doors and out into the open. This was amazing for me to watch because the disc was apparently so light weight that a team of five men was all that was required to move the vehicle out of the hanger. The disc sat on its own landing pad – a metal circular device with four legs that ended in wheels. Once moved outdoors, the wheels were locked down so the pad could not move during tests.

As I walked towards the disc, I noticed that the ground in front of the hangers appeared to be the normal rocky dessert floor. There were no sidewalks or concrete runways to indicate that this area was anything more than an arid Nevada dessert. I imagine that this attention to detail prevented the area from being spotted by overhead spy satellites. In fact, all of our tests took place when there was no chance of being spotted from those secret eyes in the sky.

My instructions were precise and to the point – I was to enter the disc and sit in the chair. After the door was closed, I was to concentrate on trying to make the vehicle lift off the pad, hover in the air and land again. There were to be no attempts to do anything else except what was instructed. To vary from the program would have strict and immediate consequences.

The chair was a rather simple affair, and void of any seat or shoulder belts. I thought this extremely odd considering what was involved. I soon learned why.

While it would be nice to say that my first attempt to control this strange machine was a rousing success – in reality the test ended in utter failure. Despite what I thought were my best efforts, I was unable to move the disc. I simply drew a blank. I just sat useless in the chair and waited for the door to be opened. If any of the scientists conducting the test were disappointed by my lack of success, it didn't show. They thanked me for my help and scheduled my return the very next day.

The next day I was ready to try again. I had spent the evening in my rented apartment in Las Vegas going over my mental exercises. These exercises were taught to me during my remote viewing training. I had found them useful in other aspects of my life, so I thought this was a perfect opportunity to put my brain through its paces.

Unfortunately, this test went no better than the first. There was no connection that I could perceive between me and the disc. It was a cold hunk of metal that was impervious to my mental commands. The scientists merely thanked me for my help and scheduled me to come back one more time.

By now I was really depressed. I knew I was no mental giant. But I had really hoped that I would be able to make this experiment work. Maybe, I thought, no one had been able to operate the disc. If it was an actual UFO from space that we had somehow managed to get a hold of, then it may operate on a science totally beyond our knowledge and capability.

Teleportation: From Star Trek To Tesla

PSI Saucers

And then it happened. It was on my fourth day and by this point I was about to give up. I was merely sitting in the chair, half asleep, waiting for my test to end for the day. Suddenly, I felt a slight tremble run throughout the ship. I snapped totally awake and looked around. Nothing had changed, everything seemed to be the same. But something was different.

There is no way for me to explain it, but the ship now seemed activated. I could just barely feel a low vibration that coursed through the body of the craft. I concentrated as hard as I could, mentally instructing the ship to lift up. But nothing happened.

I couldn't understand what was wrong. Just a second ago the ship thrummed with power. Now it just sat underneath me cold and unresponsive.

I thought back to what I was doing when the ship came alive. However, it didn't make sense, I had been daydreaming, almost asleep. I wasn't concentrating on anything, much less on making the craft move.

Then it occurred to me. It seemed so simple that I almost disregarded the thought. But now it was making sense. I had been trying to force the ship to move. Commanding it in my mind to activate and take to the sky. But it moved when my mind was relaxed and thinking of nothing in particular.

I recalled my past remote viewing instructions where I was hooked up to a biofeedback machine to learn how to put myself into an "Alpha" state. Once achieving this state of mind, the visual impressions could flow without hindrance from the conscious mind. Could it be that the ship operated on the same principle?

I relaxed myself and let my thoughts drift away. I thought of the deep, blue desert sky above me and how nice it would be to let myself go and float up into the atmosphere. At that point the ship jerked sharply upwards and came back down on its platform jarring me back into full alertness.

I had done it. I had made the ship move.

After that day it became increasingly easier to induce movement from the craft. I even managed to make it hover a full ten feet over its platform. Unfortunately, the effort was extremely draining for me. By the time a test was finished, I could barely walk unassisted out of the ship. I knew that I could not keep this up much longer.

I had thought that the scientists and technicians would have been as excited as I was by my accomplishment. But they merely asked me what my impressions were on the days activities and made notes. It was as if they had done this all before. Which in my opinion, was exactly what was going on.

My tests were obviously part of a much larger research project. I have no idea how many other test subjects were taken aboard this strange craft. And no one volunteered information.

Teleportation: From Star Trek To Tesla

The Secret of UFO Propulsion?

As a UFO pilot, I was a dismal failure. Most of the time I could only manage to rock the ship from side to side, or leap a few feet off the platform. As I stated earlier, I did manage to cause the ship to lift and hover briefly in the air, but that was the best I could do. It seemed to feed off of my energy and always left me in a weakened state.

One incident convinced me though, that the true method of propulsion for this ship goes way beyond our understanding of antigravity. It was to be my last test at Area 51.

This test seemed to be no different than the rest. I was inducing the ship to rock back and forth, when a stray thought crossed my mind. Earlier in the day I had noticed an unusual rock formation on a nearby mountain. Being an amateur rockhound, I had wondered if this formation could have been volcanic in origin.

As soon as the mental picture of the nearby mountain entered my mind, I felt the ship lurch in a way that I had not felt before. This time the ship felt like it had dropped several feet. I knew this was impossible since as far as I knew it was still seated on its platform.

Concerned, I ended the test and opened the door. Nothing appeared out of the ordinary. The ship was still on its platform in the same position as when I had earlier entered. However, the area around the platform was abuzz with technicians, and for the first time, security. I was immediately escorted back into the hanger and taken into a room to be interrogated.

I soon discovered what all the excitement was about. The questions were completely different than before. And now, wearing a uniform that I had never seen before, was a man firing questions at me alongside the scientists who were heading the project.

From their line of questioning I soon deduced that the ship had suddenly disappeared from the platform and instantly reappeared alongside the mountain that I had been thinking of at the time. Before anyone could even react, the ship just as suddenly vanished from the sky and was back on its platform, exactly as it was before. There was no time gap in between its disappearance and reappearance. It had not flown to the mountain. It was simply just one place one second and another place in the next second.

That would be my last test. After answering as best I could what I had been doing at the time of the ship's sudden departure; I was quickly flown back to Las Vegas, told to pack and sent home. That was it. No debriefing, no further questions, no answers.

My original contacts to Area 51 refused to even acknowledge that they had set up my tests in the first place. I was paid through a third party business, so there was no paper trail. It was as if the three weeks in Nevada had never happened. But it had happened. There was just no way to prove it. Since that time I have heard the same basic story from others who were contracted for short jobs at Area 51. What is really going on in those hidden hills?

Teleportation: From Star Trek To Tesla

My experience took place in the early 1980's. I have no idea how long these tests have been conducted. Nor do I have any firm evidence on why these tests took place. The stories I have heard from others, as well as some alleged insider information, leads me to consider that these were tests of a man-made craft based on the plans from a capture extraterrestrial flying saucer. However, I must add the caveat that while I was at Area 51, I was told nothing about the nature of the vehicle, so that any stories I repeat here could be disinformation to hide what is really happening at Area 51.

I have been told that the captured alien spacecraft appear to fly by the mental commands of the pilots. This seems to confirm my experiences.

The late Col. Philip Corso, whom I had the pleasure of meeting several times during our careers, wrote in his book *The Day After Roswell*, that the creatures found inside the wreaked UFO, wore skintight suits that seemed to provide a mental connection with their disc. The ship literally became an extension of the pilots, and it responded immediately to their mental commands.

The UFO's main method of operation is to use gravity waves in an extremely localized field to pull spacetime to the ship to enable it to travel from one location to another. This is instantaneous and doesn't require breaking the speed of light, because the UFO is not flying faster than light – it is pulling the location to it.

Einstein said that as an object accelerated toward the speed of light, its mass would increase, its length in the direction of travel would decrease, and its time would dilate. The rates of these changes are negligible at ordinary speeds like that of our Apollo spacecraft, but approach infinity if one approaches very close to light speed.

Some physicists interpret this to mean no object can exceed light speed. They assume that, as the mass of an object increases toward infinity, the amount of thrust (and therefore the weight of the fuel) required to accelerate it also increases toward infinity.

Einstein's theories also state that space and time are interdependent; that together they form a four-dimensional continuum; that the presence of matter warps the spacetime continuum: and that the greater the mass, the greater the curvature. Recently, we have learned that Black Holes are gravity wells created when massive stars collapse and form singularities. They warp spacetime so much that it actually turns inside-out and time runs backward.

Our scientists have deduced that the UFO builders have learned how to generate a gravity field around their craft that artificially warps spacetime. This field would, in effect, create an artificial worm-hole, a private universe that the craft could carry around. Inside, the craft would be isolated from normal spacetime. By manipulating the size, shape and strength of the field, the creatures aboard the craft may be able to fold spacetime instead of traveling through it to a distant location. This method may allow them to travel from other planets, stars, galaxies, parallel universes, and possibly even the past and future.

Teleportation: From Star Trek To Tesla

They may appear as if they are traveling through our atmosphere and performing instantaneous right angle turns at thousands of miles per hour without burning up or making any noise, materializing, disappearing, etc. But this is an illusion because time inside the field is entirely different from time outside it. Such a field would likely cause distortions not only of space and time, but of gravity, sound, light, heat, magnetism, even direction.

The only phenomenon which in any way resembles this kind of warping of space/time is teleportation, which as pointed out, is also a feature of many poltergeist cases. When an object reappears after teleportation, it flutters down like a leaf, as if it were nearly weightless, and only gradually regains its normal weight.

The same falling leaf phenomenon has been observed of flying saucers. This phenomenon has not been reported in the popular press. The falling leaf behavior may even account for the flying saucer's shape. It might make them more controllable in an atmosphere just after teleportation, before normal mass can be restored.

Physicist Bob Lazar, who also claims to have worked with captured UFOs at Area 51, says that information from briefing manuals, and his own research, shows that the discs create their own gravitational field strong enough to warp spacetime. By doing that, you're into a different mode of travel, where instead of traveling in a linear method – going from point A to B – now you can distort time and space to where you essentially bring the mountain to Mohammed; you almost bring your destination to you without moving. And since you're distorting time, all this takes place in between moments of time.

Bob Lazar says that on the bottom side of the discs are three gravity amplifiers. When they want to travel to a distant point, the disc turns on its side. The three gravity generators produce a gravitational beam. What they do is they converge the three gravity amplifier onto a point and use that as a focal point; and they bring them up to power and pull that point towards the disc. The disc itself will attach onto that point and snap back as they release space back to that point. All this happens in the distortion of time, time is not incrementing. So the speed is essentially infinite.

This is all accomplished, according to Lazar, by the use of an anti-matter reactor that uses as fuel a chip of super-heavy element 115. Element 115 sets up a gravitational field around the top and the wave guide siphons off that gravity wave, and that's channeled above the top of the disc to the lower part where there are three gravity amplifiers, which amplify and direct that gravity wave.

The low-speed mode of the discs are used during hovering, landings and take-offs. During these times the craft is unstable; it bobs and wobbles on its axis. And it's sitting on a weak gravitational field, produced by three gravity waves. The disc will often rotate to maintain stability during low-speed travel.

They essentially balance on the gravitational field that the generators put out, and they ride a "wave," like a cork does in the ocean. In that mode they're very unstable and are affected by the weather. The pilots can focus the waves behind the ship and keep falling forward and hobble around at low speed. With the second mode they increase the amplitude of the field, and the craft begins to lift, and it performs a roll maneuver. As it begins to leave the earth's gravitational field, they point the bottom of the craft at the destination. This is the second mode of travel, where they converge the three gravity amplifiers – focus them – on a point that they want to go to. Then they bring them up to full power, and this is where the tremendous time-space distortion takes place, and that whips them right to that point. This description is very reminiscent of my experience with the disc.

This propulsion system also gives rise to certain peculiar effects, including invisibility of the craft. You can be looking straight up at it, and if the gravity generators are in the proper configuration you'd just see the sky above it – you won't see the craft there. That's how there can be a group of people and only some people can be right under it and see it. It just depends how the field is bent. It's also the reason why the crafts appear as if they're making 90-degree turns at some incredible speed; it's just the time and space distortion that you're seeing. You're not seeing the actual event.

If the vehicles look like they're flying at seven thousand miles per hour and they make a right-angled turn, it's not necessarily what they're doing. They can appear that way because of the gravitational distortion. A good analogy is that you're always looking at a mirage, it's only when the craft is shut off and sitting on the ground can you see it for what it really is. This is why, when seen on the ground, that UFOs can appear to vanish instantly.

When it is operating in the air, you're just looking at a tremendously distorted image, and it will appear like it is changing shape, stopping or going, and it could be flying almost like an airplane, but it would never look that way to an observer on the ground. As the output of the gravitational amplifiers becomes more intense, the form of spacetime around the disc not only bends upward, but at maximum distortion actually folds over into almost a heart shape around the top of the disc.

This spacetime distortion is taking place 360 degrees around the disc, so if you were looking at the disc from the top, the space-time distortion would be in the shape of a doughnut. When the gravitational field around the disc is so intense, that the spacetime distortion around the disc achieves maximum distortion and is folded up into this heart shaped form, the disc can't be seen from any vantage point, and for all practical purposes is invisible. All you could see would be the sky surrounding it.

The energy transmitted within the craft is accomplished essentially without wires; it is almost a Nikola Tesla setup. It seems like each sub component on the disc are attuned to the frequency that the reactor was operating, so essentially the amplifiers themselves received the electrical energy like a Tesla coil transmits power to a fluorescent tube. That is how the amplifiers receive the power and through the waveguide to receive the basic wave. It's very similar to a microwave amplifier.

Teleportation: From Star Trek To Tesla

UFOs And Teleportation

Since my experience, I have returned to Area 51 several times to stand on a dark and lonely dirt road, watching for strange lights flying over the mountains that separated us from the mysterious base. This area eventually became filled with self-styled UFO investigators bringing carloads of tourists hoping for a glimpse of UFOs being test flown by the secret government.

In the mid 1980's I had begun to hear rumors about unusual lights that could be seen flying in the airspace above Groom Lake. This was something I had to see for myself, and I was not disappointed.

In October of 1993 I once again returned to south-central Nevada to see if UFOs were indeed flying over Dreamland. We took state highway 375 to the town of Rachel and turned onto the dirt road at marker 29 and went another four miles until the road split into three directions. This was Groom Lake road and here one could follow the road another ten miles in to the boundary of the base. We had no intention of going that far so we pulled off to the side and took a look around. The night sky was clear and the temperature was pleasant, so we settled down with a couple of fold out chairs and waited.

Around 2:00 AM we noticed over the distant mountains the sudden appearance of two orange colored lights. One second they weren't there, the next they were. They simply appeared out of nowhere. From our vantage point the lights appeared to be globe shaped and perfectly still. They almost looked like two fiery eyes looking down from above.

As we watched the motionless lights, they blinked out and blinked on again to our right. At first I thought we were looking at two different lights. Their reappearance was so sudden, and obviously miles away from their starting point that it seemed impossible for them to travel that great a distance in so short time. However, from my experience at S-4 I knew that such maneuvers were possible.

We watched the two lights for more than 45 minutes as they streaked and danced in the sky. At one point they flew so close to our camp that I was certain they knew we were there watching them. One even hovered close to the ground and about a half a mile from our location, while its companion bobbed up and down like a yo-yo on a string above it.

As I watched the two strange lights, I wondered if I was watching the same type of ship that I had attempted to fly. Their size appeared to be the same. If this was the case, a lot had happened in the years since my involvement.

From what I have gathered, some UFOs use a form of teleportation to get from place to place. Unlike the traditional concept of taking an object apart atom by atom and then broadcasting the bits to a distant receiving station, UFOs use gravity to pull space to them, an instantaneous mode of travel. And it does work, I have seen it, I was there.

CHAPTER 7
SECRET RESEARCH IN SPACETIME TRAVEL

This secret paper was discovered in 1992 by James Hartman and posted on the Keelynet website (www.keelynet.com). Hartman writes that after discussing it with others on the Internet in 1995, he felt that this information could be of great use for both engineers and other researchers interested in the subject.

Other documents by Alan C. Holt have illustrated a winged disc that has a circular center with the same glowing bright center seen on wedge-shaped UFOs sighted all over the world. It would seem that someone has developed a working model of Holt's disc.

"I further believe this type of propulsion may be what has been seen over New York's Hudson Valley and over Belgium in the past 10 years," Hartman said. "The A.C. Holt Document A80-38979: *Prospects for a breakthrough in field-dependent propulsion* is the best illustration document I have seen on this type of propulsion."

===

JSC-16073

(NASA-TM-80961)
FIELD RESONANCE PROPULSION CONCEPT (NASA)
N80-19184 13 p HC A02/MP A01 CSCL 21C
G3/20 14761 - August 1979
NASA - National Aeronautics and Space Administration
Lyndon B. Johnson Space Center

FOREWORD

The speculative "propulsion" concept described in this paper was presented at a special session of the 15th Joint AIAA/SAE/ASME Propulsion Conference (June 18-20, 1979), "Propulsion Concepts for Galactic Spacecraft". The concept was developed as the result of private, unofficial research. NASA is not involved in UFO research. However, the research which may be stimulated by this paper could result in the verification of essential elements of this concept and in feasibility studies concerning the development of a new generation of NASA spacecraft. - Alan C. Holt

Teleportation: From Star Trek To Tesla

ABSTRACT

A new propulsion concept has been developed based on a proposed resonance between coherent, pulsed electromagnetic wave forms and gravitational wave forms (or space-time metrics). Using this concept, a spacecraft "propulsion" system potentially capable of galactic and inter-galactic travel without prohibitive "travel times" has been designed. The "propulsion" system utilizes recent research associated with magnetic field line merging, hydromagnetic wave effects, free-electron lasers, laser generation of megagauss fields, and special structural and containment metals. Research required to determine potential, field resonance characteristics and to evaluate various aspects of the spacecraft "propulsion" design is described.

FIELD RESONANCE "PROPULSION" CONCEPT

ASSUMPTIONS

The field resonance "propulsion" concept has been developed utilizing recent research into causes of solar flares, magnetic substorms, black holes, quasars, and UFOs.

The concept is based on two assumptions:

(1) Spacetime is a "projection" of a higher dimensional space in much the same way that a hologram is a projection or a subset of our spacetime reality.
(2) A relationship exists between electromagnetic/hydromagnetic fields and gravitational fields - that is, Einstein's long sought for unified field theory can be developed. Mathematical relationships have been developed and theoretical concepts have been proposed to describe the causes and effects associated with the assumptions, but experimental data is required to develop the correct theoretical basis for the assumptions (Rachman and Dutheil, 1979). Specific research in a number of areas is needed and will be described later.

ASTROPHYSICAL RESEARCH

There does exist, however, some astrophysical data which tends to support these assumptions.

Teleportation: From Star Trek To Tesla

For example, astronomers have speculated that a relationship may exist between black holes and quasars (white holes). The energy and matter which leaves space-time in a black hole may reappear at a white hole at some distant spacetime point.

For this transfer of energy from one space-time point to another to occur, some type of hyperspace or higher dimensional space (4th & 5th) is required. Assumption 2 may be the cause of the large amount of energy released in solar flares.

In sunspot regions where solar flairs occur, the 2-3 thousand gauss magnetic fields are configured such that the positive and negative polarities are in close proximity with each other. Where the positive and negative magnetic field lines are nearly anti-parallel a process called magnetic field line merging can take place.

In this process the oppositely directed field lines break and re-connect expelling fields and plasma out from the sides. As a result magnetic energy is converted into kinetic energy.

The magnetic field line merging process has been proposed as the most likely explanation for solar flare eruptions. However, some flares can release energy which equals 10% of the suns' total output in a second. This large amount of energy is difficult to achieve with the magnetic field line merging concept.

Thus it may be that the configuration of the magnetic fields and associated hydromagnetic waves (oscillation of field lines) may induce a "resonance" with gravitational fields resulting in a release of gravitational as well as magnetic energy.

It is well known that the geometrical relationships of the magnetic fields (and thus the field gradients) are more important to the production of solar flairs than the magnitude of the field strength.

A strongly convoluted boundary between magnetic polarities results in a high probability for large and frequent flares. Another fact of interest is that hydromagnetic waves generated by solar flare have been observed to propagate across the chromospheric surface and trigger flares in other sun spot regions.

Alfven waves, which appear to be the dominant wave form involved, change only the geometry of the field lines. This effect also indicates that the initiation of solar flares definitely depends on geometrical relationships as do the properties of space-time and gravitational fields.

Magnetic field line merging has also been used to explain the interaction of the solar wind (and associated fields) with the Earth's magnetic fields at the magnetopause and the generation of magnetic substorms which often are triggered by solar flares.

The magnetic fields line merging process is also an essential part of the field resonance "propulsion" concept.

UFO RESEARCH

UFO studies indicate that the unexplained (residual UFO) phenomena may be due to extraterrestrial visitors, parapsychological experiences, or a combination of the two. If some UFO phenomena are caused by extraterrestrial visitors in very advanced spacecraft, then the frequency of visits and the larger number of different types of visitors (many different humanoids have been described) imply an ability to cross large stretches of galactic and inter-galactic space in relatively short time frames.

If the speed of light is a true limit of velocity in space-time, then the potential extraterrestrial visitors must utilize a form of transportation which transcends space and time to keep the trip times short.

UFOs are often observed to disappear instantaneously. In a subset of these cases, the UFO later reappears at a nearby location implying a disappearance from and a reappearance into spacetime.

The high speed, right angle turns, abrupt stops or accelerations of UFOs, and the absence of sonic booms despite calculated speeds of 22,000 mph or more suggest that UFOs may generate an artificial gravitational field or otherwise use properties of space-time which we are not familiar with.

UFO propulsion systems appear to involve electromagnetic or hydromagnetic processes as evidenced by radiative effects on the environment such as burns, dehydration, stopping of automobile engines, TV and radio disruptions, melting or alteration of ground and road surfaces, power disruptions, and static electricity effects. This data suggests that the unknown relationship between electromagnetic and gravitational fields may be used in UFO propulsion systems.

HISTORICAL DEVELOPMENT OF PROPULSION CONCEPT

The concept of transcending space-time for interstellar and intergalactic travel was felt to be a viable possibility based solely on theoretical studies in physics and astrophysics and UFO studies.

But with the postulation and potential discovery of black holes, the potential for experimental support for this concept in the near future has been established. Astronomers have already speculated on the possibility that the mass and energy absorbed by a black hole may reappear in white holes or quasars.

For mass and energy to be transported from one space-time point to another without an observable connection implies some form of higher dimensional space transcending space-time as we know it.

The existence of a higher dimensional space would open up new space travel possibilities since the connections between spacetime points through this hyperspace could involve laws much different than those which are so constraining in space-time.

For example, if each spacetime point can be assumed to be unique and if a relationship does exist between gravitational and electromagnetic fields, then each space-time point defined by a geometric form (Einstein's tensor metric) would be expected to have natural resonances with certain electro/hydro magnetic wave forms (geometric forms) which are (effectively) higher or lower harmonics of that space-time point (which is a "projection" of higher dimensional space).

The potential existence of the spacetime harmonics mean that if a craft at a specified space-time point artificially generates a configuration of electro/hydromagnetic fields which has a resonance with a distant spacetime point, a basic imbalance would be created.

This imbalance would be out of harmony with the "projection laws" which create the spacetime properties from higher dimensional properties.

Teleportation: From Star Trek To Tesla

Forces would be set in motion to re-establish a balance which in this case require that the craft and its fields be located at the spacetime point where such field configurations are a natural harmonic of that spacetime metric.

The best analogy in every day life is that of tuning a radio. The tuning knob in this case is the spacecraft's mechanism for changing the configuration of the magnetic wave form. The radio stations are various space-time points. For a radio, the signal (which is always there) is manifested through the speakers associated with the radio, for the field resonance system the speakers exist only at the radio station.

Once the spacecraft's magnetic wave form is tuned into a distant spacetime point, that wave form is forced to manifest only at that spacetime point according to the "projection laws".

A means of generating the electro-/hydromagnetic configurations which would be required remained a problem until the study of magnetic field line merging and hydromagnetic wave effects was undertaken.

With the completion of recent thesis research, a possible means for generating the magnetic wave was determined (Holt, 1979). This research along with the results of fusion research resulted in the proposed implementation of a spacecraft which generates megagauss magnetic fields which are oppositely directed.

These oppositely directed field lines will merge and re-connect expelled field lines and plasma out to the sides. In this initial concept, expelled field lines again meet oppositely directed field lines and merge again, possibly forcing large amounts of electromagnetic and hydromagnetic radiation outside the craft.

This concept is currently undergoing revision.

FIELD RESONANCE CONCEPT

In the proposed implementation high powered lasers are used to generate the megagauss fields by the effects of non-colinear temperature and density gradients (Kruer and Eastabrook, 1977; Max et al., 1978; Nishihara and Ohsawa, 1976).

By alternately pulsing adjacent laser sets, the location of the merging process can be made to oscillate back and forth at a desired rate. The amount of laser energy can be varied to alter the strength of the magnetic fields. Thus by changing the laser's power and/or wavelength and the pulse frequency, the resultant magnetic wave form can be changed or tuned to the desired harmonic of a distant spacetime point.

If flight to a nearby location is desired, the magnetic wave form can be continuously re-tuned allowing very quick and short spacetime transformations. For an observer it would appear to be a smooth flight in much the same way that we do not see the individual frames of a film.

For large jumps or travel to very distant spacetime points, some means of preventing the tuning to intermediate spacetime points is required.

In recent thesis work it was shown that hydromagnetic waves can inhibit or enhance magnetic field line merging depending on a number of factors including the magnitude and frequency of the hydromagnetic waves (Holt, 1979).

Thus by generating the appropriate hydromagnetic waves, the generation of a coherent magnetic wave form can be prevented until the laser characteristics have been adjusted so as to generate the magnetic wave form which is a harmonic of the distant spacetime point.

Then the hydromagnetic waves can be stopped and the magnetic wave form can develop. The medium in which the magnetic field line merging occurs (vacuum, ionized gas, etc), energy storage requirements, the pumping mechanism for the lasers, and crew control interfaces are currently being studied.

To avoid large energy storage requirements, magnetic pumping of the lasers may be utilized to allow a reuse of a portion of the laser energy which is otherwise radiated.

It may also be possible to obtain substantial energy during the initial part of the process of tuning to a new spacetime point through the resonance with gravitational fields.

By allowing the craft to continuously shift or oscillate to new, nearby spacetime points, a continual supply of energy could be obtained while in a hovering mode. The oscillation could be accomplished by the modulation of magnetic field line merging by adjusting hydromagnetic wave parameters. This technique would provide energy for the lasers without disrupting major laser re-tuning processes.

The outer shape of the proposed spacecraft is a "winged disc." The wings would be used for testing purposes only and could later be eliminated following sufficient testing (Grief and Tolhurst, 1963).

RESEARCH REQUIRED

The magnetic field line merging process and the effects of varying associated parameters requires extensive study. Specific laboratory tests should be conducted, some of which could be conducted in association with fusion research.

These tests should be supplemented by computer modeling studies and the collection of experimental data on magnetic field configurations and hyromagnetic wave fluctuations before and after solar flares and geomagnetic substorms.

The generation of megagauss magnetic fields by high powered lasers and the effects of varying laser parameters should be studied in laboratory experiments.

Much of this research could be accomplished in association with fusion research.

Containment and protective materials and structures for the spacecraft could also be a by-product of fusion research. The pumping of lasers by the use of magnetic fields should also be studied in laboratory experiments.

Extensive measurements of the radiation emitted by UFOs would be extremely useful in gaining insight into the physics and technology UFOs may represent.

These measurements should include photographic spectra (using special filters for cameras), magnetic measurements (using high sensitivity detectors), recording of radio frequencies, and the use of Geiger counters to detect high energy particles.

A large distribution of special detectors, as described above, should be made to UFO investigators, police departments, and other interested individuals. In addition, portable modules containing sophisticated detectors should be located at strategic positions in various countries to allow rapid transport to high activity areas. Effects on minerals, plants, animals, and humans should be studied to gain insight into radiation emitted by UFOs. Effects on humans such as burns, symptoms of radiation sickness, and other body changes should be thoroughly investigated by physicians.

Regressive hypnosis should be used extensively to obtain technologically valuable information from UFO close encounter participants. Three dimensional models of exterior and interior of UFOs should be made for cases which are significant from a technological point of view.

As soon as some of the basic interactions are better understood or after a thorough review of the state of the art of hydromagnetic interactions is completed, small scale test models of the propulsion system and spacecraft should be constructed and extensive testing performed.

The testing should proceed hand-in-hand with the results of other research. Several test models may be needed in case one or two are "lost."

FINAL COMMENTS

Since the field resonance concept utilizes two unproven assumptions and currently unverified magnetohydrodynamic processes, the concept is likely to require extensive modification and perhaps a complete revamping following initial research activity and subsequent feasibility studies.

I am confident, however, that the research accomplished under the motivation of this or a similar propulsion concept will result in theoretical and technological advances and ultimate break through in propulsion and energy systems.

===

It is obvious that the research indicated in this paper is derived from investigations into the saucers that were kept at Area 51. Clearly though the author only has a rudimentary understanding of the true nature of how the discs operate and the science involved.

Since this paper was written, one can only speculate on the advancements that have taken place in grasping the physics involved with teleportation using gravity waves to bend space/time. Based on the experiences of Bob Lazar and others, fascinating achievements in understanding the true nature of UFOs have been accomplished. Nevertheless, this field of research is still in its infancy, and there are still huge gaps in our understanding due to the inadequacies of our modern science.

CHAPTER 8
UFO ABDUCTIONS: THE TELEPORTATION FACTOR

Many individuals claim to have been taken aboard UFOs, and most tell similar stories. These abductees find themselves in a domed and circular room filled with bright light, and cold air. They are placed on an examining table where the aliens conduct a medical examination using strange probes and scanning devices.

Biological samples, such as hair, skin, and genetic material are taken. After the examination, they are often shown a three-dimensional image, usually of some very emotional situation, such as a planet devastated by war or natural disaster. The aliens seem interested in understanding human emotions. They seem to communicate by telepathy, instructing the abductees to forget what happened.

After the abduction, the witness often remembers very little, and most notice that an unusual amount of time has passed without explanation. Many experience physical or psychological symptoms indicating that something unusual has happened. These symptoms are often reported as bad dreams, depression, and family problems. Unfortunately, by most accounts of alien abductions, a person in this situation has very little control and simply may not be able to do anything to prevent what is happening.

The one phenomenon most often reported by abductees, but usually glossed over by the utter strangeness of the incident as a whole, are the reports that the abductees are usually teleported from their homes, cars, etc. and into a completely different environment.

The teleportation is often preceded by the witness being struck and enveloped by a bright light that transports them out of their familiar surroundings and into the classic abduction scenario. First is a traumatic event in which the light or some kind of energy paralyzes the person, whether they're in their home or they're driving a car. They feel themselves being removed from wherever they were. They float through a wall or out of a car, carried up or instantly teleported on this beam of light into a craft and once there subjected to a number of now familiar abduction procedures.

As reported by Dan Wright in *The Entities - Initial Findings of the Abduction Transcription Project (A MUFON Special Report),* in a study of 95 abduction cases, the subjects indicated that their own residence (generally the bedroom) was the abduction point. However, 15 subjects have also reported some other location. Ten of those were taken from their automobiles and six from a cabin or campsite.

"Whether purposeful or inadvertent, over a fourth (28%) of the subjects in the study have sensed an alien presence other than visually. This often occurred in the hours, minutes or seconds before visual confirmation. In other cases, an unseen presence was felt in the midst of a cluster of episodes but with no known close encounter on that night.

Teleportation: From Star Trek To Tesla

"The recognition was sometimes obvious - an electrical tingling, a buzzing or beeping in the person's mind, rappings and other poltergeist activity. In a few cases, the subject "just knew" that the intruders were present." All of these alleged abductions were accompanied by some form of teleportation.

In his book **Communion**, Whitley Strieber described how it felt to be teleported, and the frightening aftereffects. "Whitley ceased to exist. What was left was a body in a state of fear so great that it swept about me like a thick, suffocating curtain, turning paralysis into a condition that seemed close to death. I do not think that my ordinary humanity survived the transition."

In 1999, during the 20th Annual Rocky Mountain UFO Conference, the MUFON abductee specialist Dr. John Carpenter, who has been working with abductees since 1989, showed the crowd footage from surveillance cameras at a factory. The cameras caught the image of a worker going outside the gate and getting hit with a light from the sky. At this instant all four of the factory's security cameras went out momentarily and when they came back on, the worker had disappeared. The surveillance cameras were time encoded. This man's co-workers went out to look for him. They searched the grounds. "We couldn't find him."

One hour and fifty minutes later, according to the security cameras, the man re-appeared right where he had disappeared. He was on his knees. He vomited, and staggered away. Participants at the conference saw the proof on the security video tape. "I didn't know where I'd been – or what happened," the man explained. Dr. Carpenter reported the man felt so sick he went home. And then he became so frightened he quit work and moved away.

An Historical Overview

These types of abduction experiences are nothing new when you look at them in a historical context. Throughout recorded history there are numerous tales of strange kidnapings by non-human entities. These stories are of course couched in the sociological belief systems of their day, i.e. fairies, angels, etc. But the similarities to modern UFO abduction scenarios are fascinating, and cannot be ignored.

For example, in 1645 a teenage servant girl named Anne Jeffries of Teath, England was found barely conscious and writhing violently in the garden of the Pitt family, for whom she worked. When the girl recovered, she told her employers that she had been approached by several small, man-like creatures who touched her body and began kissing her. She felt dizzy, then she felt like she was turning into smoke and flying through the air.

When the serving girl woke up, she was in a strange, brightly lit area, surrounded by more of what she termed "the pixies." After being subjected to further study – with particular attention given to her reproductive organs, Anne once again was turned into smoke and transported back into the garden where she was eventually found.

Anne continued to undergo strange experiences for several years afterwards. Her employers reported that she would often vanish from her locked room at night, only to reappear the next day in a different part of the house. The girl blamed her disappearances on the pixies, who, she said, could take her away and return her whenever they wished.

In 1691, the Reverend Robert Kirk of Aberfoyle Scotland wrote a manuscript bearing the title: *The Secret Commonwealth of Elves, Fauns and Fairies*. This book was the first serious attempt to describe the methods and organization of the strange creatures that plagued the farmers of Scotland.

Reverend Kirk wrote that the "wee folk" were of a nature that is intermediate between man and the angels. Physically, they have very light and fluid bodies, which are comparable to a condensed cloud. They are particularly visible at dusk. They can appear and vanish at will.

Their chameleon-like bodies allow them to swim through the air, often accompanied by brightly colored, circular lights that are seen to constantly change form. The elves are fond of kidnaping people from dark lonely roads, and will even steal people right from their beds without disturbing anyone else in the household.

Those that have vanished from rooms or locked houses were thought to have been spirited away by the fairies. Most were never seen again. Those that did return, told strikingly similar tales of being transported over great distances in the twinkling of an eye. The fairy realm was described as being extremely bright, and time seemed to "stand still" for those unlucky enough to find themselves the object of attention by the little people.

Medieval occultists held the viewpoint that the fairies had a physical body, but were also able to turn into spirits that were insubstantial in nature. It was believed that the fairies used to come and speak to natural people and then vanish while one was looking at them. Fairy women used to go into houses, talk with the inhabitants, and then vanish. When they took people with them, they took body and soul together.

Today it is still a common belief in the north of Scotland that the *Sith* or fairy people existed once, and that people had to be respectful of their strange neighbors. While the Sith had no inborn antagonism towards humans, they were quick to anger and delighted in playing tricks on their mortal neighbors. Often the Sith would take a child or an adult from a household and leave them miles away from their starting point. The victim would often have no recall on how they managed to be teleported so far away in a short period of time.

These amazing feats of teleportation have been widely documented over the years. A Swedish book published in 1775 contained a legal statement, solemnly sworn on April 12, 1671, by the husband of a midwife who was taken to fairyland to assist a troll's wife in giving birth to a child. The author of the statement was a clergyman named Peter Rahm.

Teleportation: From Star Trek To Tesla

On the authority of this declaration we are called on to believe that the event recorded actually happened in the year 1660. Peter Rahm alleges that he and his wife were at their farm one evening when there came a little man, swart of face and clad in grey, who begged the declarant's wife to come and help his wife then in labour. The declarant, seeing that they had to do with a troll, prayed over his wife, blessed her, and bade her in God's name go with the stranger. She seemed to be borne along by the wind and disappeared.

The book states that the midwife came home "in the same manner," having refused any food offered to her while in the troll's company.

This story is one of many in folklore of humans who have gone to fairyland of their own will, either going to perform some service, delivering a message or taking one back. But just as often are the accounts of strange abductions by the fairies. They take men and women, especially pregnant women or young mothers, and are extremely fond of taking young children. Sometimes, they substitute a fairy child for the real one.

It was said that the fairies wanted to improve their race by carrying off young human women to impregnate with fairy seed. The women, now carrying a fairy child, would be returned home just as mysteriously as they had been taken. After awhile, the fairies would come and take the unborn child to raise themselves, or spirit it away before it reached the age of three.

It is all too obvious that these ancient stories show a fascinating similarity to modern tales of alien abductions by small, humanoid beings. The parallels become disturbing when you consider that many victims of alleged alien abductions report that they were chosen to conceive a alien/human hybrid child in order to improve the stock of extraterrestrials.

Current UFO reports along with their associated phenomena have little changed from the old tales of angels, spirits and the little people. In Song-Zi Xian, China in 1880, a farmer by the name of Jut Tan was walking home when he saw a glowing object in the bushes. He felt himself dissolving and floating away in the air, overcome by sudden paralysis. A humming noise filled the air and the farmer lost consciousness. When he later awoke on a mountain, he found that two weeks had elapsed, and he was in Guizhou Province, 300 miles from home.

One of the more unusual stories involving alien abduction and teleportation was reported by Timothy Green Beckley in the Fall 1975 issue of the now defunct *UFO Report*. On October 25, 1974, Carl Higdon of Rawlins, Wyoming was out hunting elk in a remote section of the Medicine Bow National Forest. Higdon had spotted a group of five elk and had raised and fired his rifle when something totally unexpected happened.

"I couldn't believe my senses. Instead of a powerful blast, the 7mm bullet left the gun's barrel noiselessly and in slow motion. It floated like a butterfly, finally falling to the ground about 50 feet from where I stood. I was awe-struck – I froze. All around me there was a painful silence. Not a chirping bird or the rustling of leaves on nearby trees could be heard. The only sensation I could detect was a tingling feeling which crawled up my spine. This was similar to the feeling you often get before a fierce thunderstorm, when the air is full of static electricity."

Realizing that something was terribly wrong, Higdon turned around and spotted a strange looking man standing in the nearby shadows. The being resembled a human, but with some startling differences.

"He was well over six feet in height and weighed around 180 pounds. He was dressed in a tight fitting, one-piece outfit, similar to a wet-suit scuba divers wear. Around his waist, the creature wore a thick metal belt. In the middle of this was a six-pointed star, and directly underneath the star on the belt was an unidentifiable emblem. Crisscrossing its chest were a couple of belts that looked to be a harness."

Its most unusual feature, according to the hunter, was the odd appearance of his head and face. It was unlike anything he had ever seen before.

"The creature's face ran directly into his neck. No chin was visible. His face just seemed to blend right into his throat. He had no jaw bone. Its skin was yellow, very similar to an Oriental's. It had no detectable ears, and his eyes were small and lacked eyebrows."

When the being opened his mouth, Higdon saw two sets of extremely large teeth, three on top and three below. The dome of his skull was covered with the coarsest hair imaginable. It looked as if it had golden colored straw growing out of its head. Each strand poked up a couple of inches from its scalp, and sticking out of its forehead were two antenna-like rods.

The strange creature asked Higdon if he was hungry. When the hunter said that he was, the alien waved a pointed object where his right hand should have been and floated a small packet containing four pills to the startled man.

"He told me to take one of them, explaining that it would last for four days."

Directly behind the alien, Higdon noticed the sun's rays reflecting on something in the glade. "There, not far from us, was a transparent, cube-shaped object resting on the ground. To me it looked like a huge Christmas package, flat on all sides, like a box. I couldn't see any landing gear or entrance. It was much smaller that any of our commercial or military planes. In fact, this thing couldn't have been more that five feet high, seven feet long, and four and a half feet wide. Tiny is the only word I can think of to accurately describe its size."

Realizing that Higdon had seen his ship, the being asked, "Do you want to come along?" Higdon, realizing he had little choice, reluctantly agreed. The being once again pointed the rod-like appendage where his right hand should have been.

"Before I was able to move a muscle, presto, I found myself inside this contraption. It was instantaneous. How I was able to fit inside, remains a riddle."

Teleportation: From Star Trek To Tesla

Once strapped in a seat aboard the strange craft, Higdon was able to see the forest around him. "You could see the trees and Earth below." The alien waved his mechanical "arm" at a control panel and they were launched upwards.

For a fleeting moment after lift-off, Higdon saw, through the transparent walls, the Earth drifting below. Also on board were the five elk that Higdon had intended to kill. They seemed to be in a cage, "like a corral" with "cross pieces" – bars – preventing them from moving around. "They seemed to be frozen in place," he remembered. It was as if he were gazing at some giant stuffed animals.

Higdon later recalled under hypnotic regression, conducted by Dr. Leo Sprinkle, that there was a second alien on board with them.

Sprinkle: Did they call each other by name?

Higdon: Ausso One . . . I talked to him.

Sprinkle: Ausso One? Did he appear to be the expedition's leader?

Higdon: Yeah.

Higdon noticed that at no point during the journey was he able to touch, or even get near, his kidnappers. "They were careful about maintaining at least a three or four foot distance. I am positive that they were surrounded by a force field which protected them from the Earth's foreign elements."

They offered no explanation of their means of propulsion – "after all, I'm not really qualified to understand scientific terminology," Higdon noted. They did tell him that their ship operated on the principle of "magnetic force." "These people can travel as fast they want," he said.

Perhaps their major scientific achievement, which puts their technology far beyond ours, is the means that they use to move from one location to another, instantaneously. This feat seems to be accomplished by aiming their right "arm" in the direction they wish to travel. Ausso One used the term "gun" when referring to his apparatus. "He used his right arm to point where he wants to go, and he just goes. He moves freely, anywhere we want to go." Higdon believes he entered and left the ship by means of teleportation.

Following a flight that seemed to him to be about 30 minutes, a sparkling sphere loomed before them. "We landed where the lights were brightest," Carl said. "It must have been night because the lights were in a confined area – casting outward maybe 100 yards in a circle," referring to the landing pad. The lighting, he said, was intense and "artificial."

Higdon remembers that the lights were extremely bright and painful. He tried to use his hands to block out the rays of the powerful lights. "I vaguely remember shouting: 'Shut them off, they're burning me!'" Ausso One explained that they lived under a different sun and that the Earth's sun burnt them as well. "That's the reason they operate at night or stay in the shadows during the day; the light from our sun hurts them."

Teleportation: From Star Trek To Tesla

Higdon was once again instantly teleported off of the ship and into a cubicle type room where he was placed before a glass-like screen and given what he thinks was a medical examination. After the examination was completed, Ausso One told Higdon that they were taking him back home.

"He said to me that there was no further reason to detain me because I'm not any good for what they need," Carl said. "This may sound stupid, but nine years before, I had a vasectomy. Maybe this is what he meant when he said that I wasn't any good."

Carl Higdon was then briefed on the main reason the aliens were coming to Earth. "They're coming after food. Exploring, hunting, fishing. Ausso One just kept talking about hunting and meat. Their concentrated food is not enough to sustain them. They need meat to help them survive. They're coming to Earth to hunt and take back living animals to breed on their planet."

Higdon was then ushered by Ausso One, through a long corridor. "We walked back down a hall. The door opened and we stopped on a kind of platform and went down to where the spacecraft was parked."

The creature once again pointed his hand and Higdon suddenly found himself back on board. "It was like, POOF, and I was instantly back in the chair."

Carl said that he could see through the transparent walls of the ship a group of five people talking amongst themselves outside. "These people were definitely earthlings," he said. "There were three adults and two kids. The two kids, a boy and a girl, couldn't have been more than 17 or 18 years old. They didn't act like they were surprised or afraid. They just stood there like they were having a normal conversation."

Before Carl could ask his host about the people, the craft quickly took off and returned to Earth. Carl could now see that they were hovering over his parked truck. "When we got above the trees, Ausso One aimed his 'arm' at my pickup and it just disappeared into thin air."

Although he had no way of knowing it then, Higdon's truck had been instantaneously transported a distance of five miles. It had been lifted from its parking spot on the grassy knoll in McCarthy Canyon, and dropped in an area where no vehicle could possibly maneuver.

Before Ausso One teleported him to the ground, Higdon's magnum rifle was returned. Resting next to the seat, on the floor of the cubicle, the gun suddenly levitated into his hands. Then suddenly, Higdon was back on the ground. "I don't know how it happened. I just was gone from my seat to the ground, just like that."

Higdon was understandably in shock over his ordeal, and was subsequently found hours later by rescuers. He was discovered next to his mud-bound truck, delirious and still complaining about the bright lights that hurt his eyes. He could offer no explanation on how he had managed to get his truck to such an inhospitable location.

Teleportation: From Star Trek To Tesla

Unconventional Means

Carl Higdon's story is a refreshing change from the more recent alien abduction tales. Most, if not all, have become repetitive to the point of boredom in their similarities. These similarities may also include incidents of teleportation, or at least methods of collecting the human abductees that seem to defy conventional means.

UFO investigator Don Worley of Indiana uncovered this interesting tale of alien abduction that allegedly happened in Nebraska in 1955. A girl identified only as "Jennie" was in her bedroom one night in October when she noticed a strange light playing against the glass of the window. When she drew back the curtain, she was horrified to see a "tiny man" floating just outside the window.

"He was tinier than four feet, and he had on a tight white cap, like a swimmers cap. The head was shaped like an egg, and the face was waxy . . . real pale, pinkish, almost greyish. It's like if you touched him you'd hurt him. The nose was just a tiny bump, like two black slits. The mouth was just a slit."

The strange entity telepathically wills Jennie to come closer to the window. "I really don't want to listen to him," she says. The creature floated away from the house and towards a glowing object shaped like two dessert bowls placed together. The object was floating quietly in the air over the backyard of the house.

In her nightgown, Jennie floats out towards the UFO. She can't explain how it happened, but she remembers seeing dirt and cobwebs inside the house wall as she passes right through it and into the night air. "I felt like I had become a ghost, like my body no longer had any substance. I floated right through the wall of my bedroom, feeling every layer in the wall. I could feel the paint, the plaster underneath, the wood, and finally the outside surface of the wall. I just passed right through it like it wasn't even there."

As Jennie floats towards the UFO, she notices that it seems to fade in and out of existence, periodically allowing the yard below to be seen. "When it goes, you can almost see inside it. First it's there, then it isn't." As with the wall of her house, Jennie floats right through the wall of the UFO.

Inside the UFO, Jennie undergoes the typical medical examination. She is clamped onto a table and blood is sucked up through a little tube. "I kept saying that it hurts and he's pretending he doesn't care. But he cares. He's smiling because he knows I know it really doesn't hurt. I only think it does."

Afterwards Jennie is returned to her room in the same manner in which she left, and the UFO vanishes. The next morning, when Jennie wakes, she remembers her strange experience. "I looked out my window and the elm tree is burnt. Pop says it was hit by lightning. But it wasn't."

Teleportation: From Star Trek To Tesla

Ivan Boyes of Auckland, Canada claims that he was "beamed aboard" a spacecraft in 1960. He recounted his tale in the May, 1977 issue of *Official UFO Special*. At the time, Ivan was a 15-year-old student who had never even heard of UFOs.

On the evening of October 19, Ivan had been home doing school work until a strange feeling overcame him. "I had the overpowering urge to get up and go outside. The feeling soon became a compulsion. I went out around seven o'clock and walked in a southerly direction to the Queen Elizabeth highway. I was being pulled by an unknown power to an equally unknown destination."

Ivan soon hitched a ride out toward the Welland area of Ontario, where he found himself walking into the bleak countryside to the north of the highway. "Suddenly the whole area around me was lit up by a brilliant light that was just above me. As I looked up toward the light, I saw within the light a large, bluish-white craft. It was oval-shaped and must have been at least 100 feet above the ground and at least that much in diameter."

In a fit of courage, Ivan called out to the strange craft hovering silently overhead. "Who are you? What do you want?"

Then the unexpected happened. A loud resonating voice spoke out from the flying saucer saying: "Do not be afraid. We will not harm you. Just relax and prepare yourself."

Before the boy could react he was plunged into darkness. "The next thing I could remember was suddenly finding myself on board what I took to be the flying saucer. Looking around from the inside of some form of transparent tube that I had found myself standing in, I noticed that the craft was circular in design and made up of an opaque whitish-blue colored metal. The instrument panels and the floor were of the same texture and metal. Close by stood a man six feet in height with snow-white hair."

There were two other aliens of the same composure on board the saucer, except that their hair was of an ebony color. "The alien then approached me saying: 'We have brought you here because there are many important things to be done in the future of the Earth; these will change its history for the better or for the worse. It all depends on where a person stands in relation to his civilization.'"

The alien showed Ivan a screen that displayed a television image of the coast of Brazil being wracked by earthquakes, fires and finally a tidal wave which destroyed the city of Rio de Janeiro. He was then told that this was what was in store for the entire planet in the near future. "Our civilization, like countless others through the ages, was soon to perish in an terrifying cataclysm."

It was now time to leave, and the alien escorted Boyes over to several transparent tubes. These were the same devices in which he had originally arrived. The front part of the tubes were opened and Boyes stepped in. The aliens made some adjustments on a nearby instrument panel closing the transparent cover of the tube. Ivan now stood in complete silence watching the tube as it began to energize.

Teleportation: From Star Trek To Tesla

"Swirls of vivid colors permeated me as I was transformed into energy." How many people can describe the experience of spirit and soul transference and that of teleportation? "Strange, vivid colors surrounded me after the hazy, yellowish hues of the first process of energization had begun.

"The colors appeared as swirling indigo colors: whites, blues, yellows and many other indescribable colors. Waves of caressing energy ran through my body, purging my spirit, rendering my body a renovated house for a new resident.

"It was the most omnipresent feeling of well-being that a person could experience, not only the experience of feeling and seeing colors but also the wisdom and language it possessed. The whole of the universe seemed to be in those colors, it would take only time and meditation to understand the laws and secrets it talked about. How long I was in this state I really do not know, but it seemed an eternity."

Ivan Boyes awoke to find himself flat on the ground and barely able to move. It was now 7:30 in the morning and the UFO was no longer in sight. Unlike other abduction stories, Ivan fully remembered his ordeal, and felt gratitude that he had been chosen to be picked up by the aliens. Ivan claimed continued contacts with the aliens, who said that in the distant past, they were once inhabitants of Earth. Like other contactee stories, Ivan Boyes' tales grew more fantastic with each event as the flying saucer people sent him on time travel missions into the past.

Another unusual UFO incident occurred in Bahia Blanca, Argentina. Twenty-eight-year-old Carlos Alberto Diaz was returning from a part-time night job as a waiter providing extra money for his family. It was around 4:00 AM on January 5, 1975, when suddenly there was an intense hum that split the quiet morning. Carlos was blinded and paralyzed by a beam of light that shot down out of the sky. Unable to flee, Carlos felt himself being drawn away in the light beam. He lost consciousness to awake inside a strange room, with air coming through holes in the floor.

As he tried to move away from them, he was approached by a being that was normal in height and covered by a suit and helmet that hid any details. The being used a suction devise to take pieces of the young man's hair. When Carlos touched the entity, it had a texture like rubber.

Carlos remembers that he was once again hit by an intense beam of white light that seemed to absorb the man into itself. He awoke back on the ground to find that it was now 8:00AM and he was close to the city of Retiro, over 300 miles from where the light first struck him.

Carlos Diaz refused all suggestions of publicity, but he would happily tell his story to investigators for free. He said later: "I don't know if you will believe me. If someone were to tell me, I surely wouldn't. The only thing I know is that this happened to me."

In 1978, after a wave of flying-saucer sightings in the Gisborne, New Zealand area, three young women went UFO hunting. Although they returned home with no conscious memories of seeing anything odd, their log indicated an unexplained two-hour gap.

When one of the women later underwent hypnotic regression, she recalled being abducted by aliens along with one of her friends. The victim recalled being turned into "nothingness" and "sucked up" by a beam of light and finding herself and her friend laid out on slabs for examination by aliens, described as having long, thin faces with large black eyes.

In 1982, a twelve-year-old Malaysian girl named Maswati Pilus was going to the river behind her house to wash some clothes. Suddenly she was confronted by a female being about her own size. All the sounds of the village nearby disappeared, and it seemed as if only she and the entity existed.

The creature had pale white skin and hair, and her fingers were longer than a human's. She was dressed in a white outfit that covered her entire body. The strange entity invited Maswati to come with her and see her home land. Maswati reluctantly agreed to go along.

The little girl felt no fear as she reached out to take the hand of her new acquaintance. When she did, she suddenly found that she was no longer in her village, but was now in a "bright and beautiful place." It seemed to her that time was whizzing by, and she soon lost consciousness.

Maswati was discovered two days later by frantic relatives on the ground, unconscious, not far from her home. She was discovered in a location that had been searched several times before. As with other cases, the little girl could offer no explanation on where she had been, or how she had mysteriously appeared in a spot previously searched.

Incident Over The Brooklyn Bridge

Incredible stories by those who say they were contacted by extraterrestrials have become somewhat commonplace in the journals of UFO lore. Most of these experiences have to be taken for what they are worth due to the lack of any evidence or eyewitnesses. However, there is one controversial case where a woman claimed to have been teleported out of her bedroom by aliens, in full view of a number of amazed onlookers.

The 1989 abduction of Linda Cortile was investigated by abduction researcher Budd Hopkins, who detailed this amazing story in his book *Witnessed, The True Story of the Brooklyn Bridge UFO Abductions*. Linda Cortile reported to Hopkins that on the night of November 30, 1989, she had been floated on a blue beam of light from her 12th floor apartment, located opposite the busy *New York Times* loading bay, into a nearby hovering UFO.

Over a year later Hopkins received a letter from two men, Richard and Dan who claimed to have been security officers escorting United Nations Secretary Perez de Cuellar across the city when their car stalled. Standing outside their vehicles, the witnesses looked up to see Linda and three small figures as they floated up into a beam of blue light, emanating from the underside of a large orange glowing object which then flew into the East River and disappeared.

The two eyewitnesses at first, purported themselves to be police officers and, later, admitted that they were security intelligence agents with a major government organization, which Hopkins suspected was the National Security Agency. At the time the incident occurred, Hopkins noted that there was a great deal of political unrest on the planet. "The Soviet Union was breaking up, there were riots in Czechoslovakia, this was the day that Gorbachev met with the Pope, it was the day before he met with Bush. Things were really hot."

The UFO incident occurred as the men were driving in a motorcade with several cars and political figures, returning from a late-night meeting at the United Nations. Their car died at that exact spot where the incident occurred. The men witnessed the abduction, then watched as the craft flew over their car, over the Brooklyn Bridge, and into the river.

These people later came to understand that they had also been abducted that same night. Hopkins said the main point is that we now have eye witnesses to an abduction, "which was not a concealed abduction, as they usually are," he noted, "but one that was obviously performed almost as a theater piece for the people involved."

Months later another woman came forward and claimed to have witnessed the event from the Brooklyn Bridge. She was some distance away, but saw the lights from the huge craft and then the small figures floating out of the window. She thought that someone must be making a movie with special effects, but soon realized that what she was seeing was no movie.

This story has also taken many strange twists and turns with the same bends in reality that plagued those who claimed contact with the fairy folk in centuries past. If these incidents are not somehow related, then all must have gone to the same university for a masters in mysterious activity.

If we accept that these stories are true, as the witnesses remember, then teleportation and other forms of mysterious transportation are somehow possible. Fairy folklore says that the little people have the power to spirit people a vast distance away in the twinkling of an eye. Modern tales of UFOs often recount incidents of teleportation that end with the poor victims left stranded miles away from their starting point. Different stories, possibly different methods, yet both have the same end result, teleportation.

So teleportation may not just be the wild dreams of science fiction writers. There must then be some scientific means, whether natural occurrence, mind power or fantastic device, that can make teleportation possible. The problem is that our current understanding of the universe and physics is bereft of any practical solutions to how teleportation can be accomplished. Possibly we are just not looking in the right places for answers.

CHAPTER 9
THE WEIRD WORLD OF TELEPORTATION

Almost all the data associated with UFOs have their analogies in spontaneous psychic phenomena, or have been noted to occur in the seance room.
Berthold Schwarz, *UFO Dynamics:*
Psychiatric and Psychic Aspects of the UFO Syndrome

As we have seen in previous chapters, cases of teleportation are frequently reported in association with paraphysical events, as well as UFO incidents. Often, UFO experiencers will also become targets of poltergeists and other so-called "supernatural" activity. This activity usually involves strange noises, household furniture being moved around by unseen hands, items disappearing and/or reappearing, and the sightings of apparitional figures.

UFO literature is crammed full of fascinating stories of high weirdness that would be right at home in most popular ghost books. If UFOs are spacecraft being flown by visiting extraterrestrials, why is there so much paranormal activity, including teleportation, involved with UFOs? You would think that if an alien could travel hundreds of light years to get to Earth, it would have something better to do then haunt eyewitnesses and knock their furniture around. What is the connection that we seem to be missing?

The Men-In-Black and Other Strange Beings

In the weird world of UFOs, the Men-In-Black seem to represent the devil made flesh. Mysterious men who arrive to warn, harass, even threaten UFO experiencers into silence. Stories of the MIB have become ingrained in UFO folklore to such a extent that they have become archetypical symbols of the dark side of UFOlogy.

To those who have actually been unfortunate enough to have received a visit from the MIB, the event is usually so unusual and frightening that the reality of the experience cannot be questioned. Again we are left with stories that seem to transcend our normal reality and ventures into a strange world populated by aliens, ghosts, monsters and other strange creatures. Creatures that are very real to those who happen to meet up with them.

One of the most noteworthy MIB cases on record was that centered around Dr Herbert Hopkins. In September 1976, Dr Herbert Hopkins, a 58 year-old doctor and hypnotist, was acting as consultant on an alleged UFO teleportation case in Maine. On September 11, when his wife and children had gone out leaving him alone, the telephone rang and a man identifying himself as vice-president of the New Jersey UFO Research Organization asked if he might visit Dr Hopkins that evening to discuss certain details of the case.

Teleportation: From Star Trek To Tesla

Dr Hopkins agreed seeing no reason why he should not help out a fellow investigator. He went to the back door to switch on the light so that his visitor would be able to find his way from the parking lot, but while he was there, he noticed the man already climbing the porch steps.

"I saw no car, and even if he did have a car, he could not have possibly gotten to my house that quickly from any phone," Hopkins later commented to investigators.

At the time, Dr Hopkins felt no particular surprise as he admitted his visitor. The man was dressed in a black suit, with black hat, tie and shoes, and a white shirt. "I thought, he looked like an undertaker," Hopkins later said. "His clothes were immaculate – suit unwrinkled, trousers sharply creased. When he took off his hat, he revealed himself as completely hairless, not only bald but without eyebrows or eyelashes. His skin was dead white, his lips bright red."

In the course of their conversation, Hopkins strange visitor happened to brush his lips with his grey suede gloves, and the doctor was astonished to see that his lips were smeared and that the gloves were stained with lipstick. It was only afterwards, however, that Dr Hopkins reflected further on the strangeness of his visitor's appearance and behavior.

Particularly odd was the fact that his visitor stated that his host had two coins in his pocket. It was indeed the case. He then asked the doctor to put one of the coins in his hand and to watch the coin, not himself. As Hopkins watched, the penny turned a silver color, then blue, and seemed to go out of focus, and gradually dematerialize until it vanished.

"Neither you nor anyone else on this plane will ever see that coin again," the visitor told him.

After talking a little while longer on general UFO topics, Dr Hopkins guest told him to stop all UFO investigations, especially the Maine teleportation incident, and destroy all of his recordings. At this point Dr Hopkins suddenly noticed that the visitor's speech was slowing down. The man then rose unsteadily to his feet and said, very slowly: "My energy is running low - must go now - goodbye." He walked falteringly to the door and descended the outside steps uncertainly, one at a time.

Dr Hopkins saw a bright light shining in the driveway, bluish-white and distinctly brighter than a normal car lamp. At the time, however, he assumed it must be the stranger's car, although he neither saw nor heard any vehicle.

When his family returned, they and Dr Hopkins examined the drive and found marks, but in the center of the driveway where a car's tires could not have been. By the next day, although the drive had remained undisturbed in the meantime, the marks had disappeared.

Dr Hopkins was by now feeling very disturbed about the previous night's events, especially when he recalled his visitor's bizarre behavior and actions. He was afraid, and erased the tapes of the hypnotic sessions he had been conducting, as his visitor had told him to do. He refused to have anything more to do with the Maine case.

Teleportation: From Star Trek To Tesla

Vanishing MIB – Ghosts or Teleportation?

At the time of the incidents in 1980, Maria Korn was a 14 year-old boarder at the Convent of Jesus and Mary in Milton Keynes, Buckinghamshire, Great Britain. As reported by researcher Robert Bull, Maria was receiving psychological counseling from a Dr Black for acute anorexia and sleeping difficulties.

One night Maria couldn't sleep and got up, at about 1:00 AM to look at the stars. She stood by the window overlooking the tennis court and was surprised to see a large, ball-shaped flashing light inside the tennis court itself, which was completely surrounded by wire netting. She looked at it for about five minutes, then turned away. When she looked again at about 1:30 AM the object had gone, but she heard a strange whirring sound and, looking up, she saw the object again just above the window, but there were no lights on it this time. The object then moved off rapidly.

The next morning, none of the other girls mentioned having seen anything strange during the night, and Maria kept her experience to herself. Later that morning she was playing tennis when she slipped and fell, and was surprised to see a large, shallow, circular depression seemingly burnt onto the tennis court, exactly where she had seen the object the night before. The next day the police arrived to inspect the damage, but Maria didn't tell anyone what she had seen two nights before.

Three or four months later Maria was in a math class when one of the nuns, Sister Jennifer, interrupted the class and took her out, saying that she had a visitor. Maria thought at the time that it was unusual for the class to be interrupted, visitors normally came only at weekends, and she was worried in case the visitor was bringing bad news.

Sister Jennifer showed her into a dining room where, sitting at a large table, was not one visitor but two men dressed in black. The heat was on and the room felt warm as Maria entered it, but as soon as she saw the two men she felt cold, and she put on the jacket that she was carrying.

Maria had never seen her visitors before, and she asked them who they were. They told her they were with Dr Black, her psychiatrist. She stared at their eyes, which weren't brown or blue, but a strange brownish greyish color. She found the experience frightening, and looked away.

The men asked her how she was and how she was doing at school, just a normal, polite conversation. As they talked Maria noticed several other strange things about her visitors.

They looked the same, as if they were identical twins, their skin was smooth and featureless, no beard and no sign that they ever needed to shave at all. Their hair was shiny black, each brushed in the same style and not a hair out of place, their black suits, seemingly brand new, fitted perfectly as if tailor-made and having razor-sharp creases. Their ties and

Above: MIB photographed in New Jersey by James Moseley and Tim Beckley in front of the house of John and Mary Robinson in 1968.

Below: Artwork of a MIB by Carol Ann Rodriguez.

socks were also black, exactly the same black as their trousers and jackets.

One of the men then asked Maria if anything strange had happened at the convent recently. She instantly thought that they must mean what she had seen at the tennis court – but she was afraid to admit anything she knew to the strangers, so she said no.

The two men obviously knew she was lying as they continued to press the question. Just then the school lunchtime bell rang. The man asked what the bell was for, then told Maria she had better go to lunch, adding that they had to leave.

Although the men looked to be like businessmen, they didn't appear to be wearing watches as one of them asked Maria the time. The men had been drinking coffee during their talk with Maria, but when she shook hands with them as she was saying goodbye she noticed that their hands were ice cold, despite the fact that they had been holding hot cups of coffee during most of the conversation.

One of the men asked Sister Jennifer if Maria could show them the way out, saying finally that "we'll be back to see you again." She walked to the doorway and was amazed to see that the two men were walking and swinging their arms exactly in step with each other.

They walked outside to their waiting black car, Maria noticing as they did so that, although it was a windy day, their hair didn't move, as if it was glued down.

The car had a chauffeur, also dressed in black, who must have been waiting all this time. As the car moved off Maria noticed that the number plate had white characters on a black background, which she thought strange. It also had mirror windows; Maria couldn't see through the windows into the car, but she thought that the men could see out. She caught a glimpse into the car when one of the men opened a door, but all she could see was black, no seats, no dashboard. It was as if she was looking into a black hole.

The car moved off silently. There was no sound of an engine being started, no exhaust fumes. (An electric car?) She was also mystified that although she saw the car turn out of the convent gate, she didn't see it moving up the hill that lead to the convent. It had completely disappeared on the only road leading to the convent.

Maria stood where she was for several minutes, unable to move. She could hear Sister Jennifer calling to her, but she couldn't turn around and go to her. Sister Jennifer asked her if she was alright, at which point she "snapped out of it" and was able to move again.

She asked Maria who the men were and Maria replied, saying that they were from Dr Black, although they were both surprised that Dr Black hadn't warned them that the men would be coming. When, a few weeks later, Maria saw Dr Black, she asked her about her visitors. Dr Black said she hadn't sent the men, and would never send anyone without informing Maria first.

Although Maria never saw her MIBs again, she did begin to develop psychic powers and have extraordinary experiences. While still at the convent she: Bent a spoon, Uri Geller style – Timed herself swimming under water for 5 minutes before surfacing – She had an

Teleportation: From Star Trek To Tesla

out-of-body experience. Experienced an upsurge in her creative and academic abilities - tested and verified by Dr Black. Most amazingly of all, she also claimed that on one occasion, at night, she went out into the convent yard and began to fly. This was witnessed by several other girls, who ran around trying to catch her when she flew low enough.

After Maria left the convent her strange experiences and abilities continued. She found that she could make things disappear and reappear by just thinking about them. She could make light bulbs burn out, she started a stalled car engine (with a jammed starter motor and a totally dead battery) just by willing it. She claimed to be able to turn red traffic lights to green, repeatedly, even after they had just changed to red.

Strangely enough, Maria also experienced several episodes where she was unexpectedly teleported over the distance of several miles. Once, she stepped out of her front door to get the mail and found herself standing on the sidewalk in a neighborhood over five miles from her home. This happened without any warning and, as far as she could tell, was instantaneous.

Marie can offer no explanation on why these strange events have centered on her. Many of the things that happened to her were witnessed by others and were undoubtedly real. The UFO she saw left a depression and burn marks on the ground which was there for all to see; her MIBs were seen by other people and they drank their coffee; her spoon bending, the increase in her academic and artistic abilities, light-bulb popping, some teleportation incidents, UFO sightings and other events can all be testified to.

In another weird case, also in Great Britain, Mrs. Evans (pseudonym) of Portsmouth, Hampshire went to visit the local grocer's store one morning in the autumn of 1977. In the shop she saw a tall man, dressed in black. He was ahead of her, so she stood back to wait until he was served.

As she stood in line Mrs. Evans noticed that the man's gaze was fixed upon her. She found this unnerving; he looked at her as if he knew her, as if he had been expecting her. He left the store when it was her turn at the cash register.

When she left the shop, Mrs Evans noticed that the man was standing nearby, as if waiting for someone. As she started to walk, he began walking also, keeping five or six paces ahead of her. As she watched him she began to form the impression that he was "unusual," although she couldn't say just why.

Mrs Evans was half-way home when he turned left into a side road. As she crossed the road, she looked to her left out of curiosity only to see the same person standing in the middle of the road. Facing her now, their eyes met once more. He nodded three times, without any change in his facial expression. His gaze was intense and penetrating.

Then, to Mrs Evans's utter amazement, he vanished without moving from the spot, "like someone turning out a light." Thoroughly unnerved by now, she hurried home later recalling several strange things about the man:

Teleportation: From Star Trek To Tesla

❑ His clothes looked brand new, as though they had only just been bought. He was dressed from head to foot in black (except for his white shirt).

❑ His skin was albino-type white, as was his hair, which was wispy. His eyes were jet black, and he appeared to be in his early 50s, but there were no wrinkles on his skin, and no sign of any facial hair or stubble.

❑ He had unusually broad shoulders, a narrow waist, and he walked upright with a stiff gait. There seemed to be no natural curve to his spine, which was seemingly perfectly straight.

This was not Mrs Evans only encounter with the unusual; in 1979, over a year after her original MIB encounter. In her kitchen one day, she became aware that there was a figure standing beside her. Her husband walked in and shouted "Who's that? What's he doing here?" Whereupon the figure, which did not seem to be totally solid, ran out of the open kitchen door.

On another occasion, Mrs Evans was returning home one evening with the family dog, when she saw, in the light from a street lamp, a tall figure. The figure was completely black and seemed to be wearing some kind of helmet, making her think of a scuba diver. At this moment her husband walked out of the front door and again shouted "Who's that? What's he doing here?" He was convinced by now that this was Mrs Evans's secret lover.

As her husband shouted, and as the dog started to bark, the figure glided forward, going through her neighbor's front garden hedge. She later recalled that the figure appeared at first to be completely solid and real, but it grew more transparent and eventually vanished as it slowly went through the bushes.

On another day, in broad daylight, Mrs Evans encountered a "strange little man" who appeared in front of her and walked on by. On turning around, expecting to see his back, she was shocked to find that he had vanished.

She described the man as small, about five feet tall, olive skinned, large, round dark eyes, and black hair, slicked straight back. He seemed to be wearing some kind of RAF uniform except that it looked made-to-measure, perfectly cut and stitched. His shoes looked brand new but were not the current fashion. He walked towards her with his arms held up in front of him, gazing straight ahead with blank eyes.

On yet another occasion, again in daylight, Mrs Evans was out walking, and she came upon a small van parked in the road. The van was white, with what looked like blue clouds and flowers painted on it. As she approached the van, its door suddenly opened and a little person jumped down in front of her. She just kept slowly walking as she and the little person gazed at each other.

Teleportation: From Star Trek To Tesla

At first she thought she was seeing a doll, or puppet wearing a checkered shirt and blue jeans, but she was startled as she looked into his eyes, which were jet black, marble-like, with two white dots where the pupils should have been.

He seemed to have Eskimo features, with dull black, dead straight hair, roughly cut in a page boy style. He seemed to have a knowing look in his eyes, which disturbed her. As he passed by her she tried to look over her shoulder to see him, but her neck did not seem to be moving normally and she could only see him out of the corner of her eye.

Mrs Evans thought he seemed to be a freak of some kind, although he was perfectly proportioned. In this case and others, it was the eyes which made her realize that she was not seeing something normal.

Another encounter came one evening in Mrs Evans's front garden. She noticed movement within a large bush which was in the garden, as if a cat or other small animal was inside it. She remembers that everything seemed unnaturally still and quiet.

Slowly, the bush began to part in two or three places. Instead of the cat that she was expecting, she was amazed to see faces, seemingly those of children. When she realized that these were not children, she froze, and the hairs on the back of her neck stood up on end. She began to hear soft clucking sounds, the sort of noise one makes when trying to make friends with a shy animal.

What she saw made her think of elves, pixies and the like. They did not seem to be totally solid looking and the bushes covered their lower bodies. As their misshapen hands extended out towards her she decided that she had seen enough and ran indoors.

As she ran upstairs her husband called out to her, asking if she had seen what it was that had just bolted out of the front garden. Later, with her heart still fluttering, she came down and looked nervously outside. The "elves" were gone, but she saw, walking towards her down the path near her home, a dark figure lit by an aura that moved with it. Before it suddenly disappeared, Mrs Evans saw that the figure had short, dark curly hair and a pointy face with very high cheekbones. Its eyes seemed to glow as it stared at her in a menacing way.

Like others who have had similar experiences, Mrs Evans found herself undergoing a whole series of unexplainable occurrences. As a young girl in 1947, there was a poltergeist in the family home, although she herself did not realize this and her parents who had never even heard of the poltergeist phenomenon did not tell her about it until many years later.

October 16, 1973 - her father sees a massive UFO. Winter 1977/Spring 1978 - Mrs Evans sees (with her husband) her first UFO. Christmas 1978 - early new year 1979 - UFOs, hauntings, poltergeists. Her husband and neighbors also experience these phenomena, but her house seems to be the focus. She begins to notice strange marks, burns, bruises, and puncture marks on her skin, which seem to appear in the mornings after restless nights. Mrs Evans reports seeing "about a dozen" UFOs from Christmas 1978 to November 1979.

Teleportation: From Star Trek To Tesla

A blood-like substance appeared "out of thin air" at her home, also a "transparent, jelly-like" substance. Strong smells – she, her husband and her neighbor saw a small, yellowish cloud, accompanied by a strong smell of sulphur. On another occasion there was a strong, "overpowering" smell of incense, also smells of zoo animals' cages and wet animal fur. On two occasions when "something unusual" passed over her head, she felt a click, or tap, on her temples, somewhat like a tiny electric shock.

Things appeared and disappeared in her home: keys, jewelry, eye glasses. Her purse rose from the table, flew through the air and landed in her left hand. The tea kettle whistled as if boiling, but there was no water in it, and the gas was not turned on. Flames came out of fingernails, which turned bright turquoise overnight, the color being on the underside of the nail.

A "paper tape streamer" appeared out of nowhere and flew through the air in her living room. It bore the words: "Don't be afraid - we are coming back in October." Her milkman saw her standing at her front door and waved to her, then he turned around and saw her walking down the street towards him. Other people reported seeing her at various other places when she was actually miles away.

In an article titled: *MIB Activity Reported From Victoria B.C.*, (**Flying Saucer Review**, January 1982) Dr. P.M.H. Edwards, formerly professor of linguistics at the University of Victoria, detailed an unusual case of MIB teleportation.

In October 1981, three days after a UFO sighting in Victoria, British Columbia, Grant Breiland saw two sun-tanned expressionless men who lacked fingernails observing him at a K-Mart department store. They were stiff, seemingly "at attention," and were dressed in very dark blue clothes.

They approached, and one asked Breiland in a monotonous, mechanical voice, "What is your name?" Their lips did not move when they spoke. Breiland said, "I'm not going to tell you that." The other man asked where Breiland lived, and then, "What is your number?"

Breiland did not respond. The two strange men stared at him for a few seconds, then turned and left, but Breiland followed them out of curiosity. The two men waited at the edge of a muddy, plowed field and, as Breiland watched them, he thought he heard someone call his name. Turning around, he saw no one. The two men walked into the field, and again, Breiland thought that he heard someone call out his name. Suddenly, the two men vanished three-quarters of the way across the field. Checking around, Breiland was unable to locate any footprints in the muddy field. It was as if the two men were ghosts.

Breiland noted that, mysteriously, no other persons were in sight at the busy shopping area during the entire incident, and the setting was only repopulated after the strange men had vanished. The "depopulation" anomaly has been noted in other UFO and MIB cases, and has been termed the "Oz effect" by British UFO researcher Jenny Randles. This zone of unreality seems to indicate that these incidents could be paranormal in nature.

Teleportation: From Star Trek To Tesla

Poltergeists From Outer Space

There seems to be a fine line between certain UFO experiences and classic hauntings. That is not to say that all UFO sightings will lead to paranormal experiences. However, based on a number of reliable reports, there does seem to be a certain aspect of the UFO phenomenon that either closely parallels psychic occurrences, or originates from the same source.

In other words: Some UFO incidents can be considered the same as hauntings and other paranormal events – with the same sorts of phenomena, such as: ghostly figures and visits from seemingly physical, yet bizarre people, unexplained noises, teleportations, the psychokinetic moving of objects and furniture, and even possession.

One such case that blurs the line between UFOs and the paranormal was investigated by Timothy Green Beckley and Brad Steiger. Brian Scott claimed to have been abducted by extraterrestrials and taken to a secret underground base within the Superstition Mountains of Arizona.

The Superstition Mountains has a long and colorful history of unusual events and hauntings. There have been several reported cases where strange voices have driven treasure-hunters insane and, in some cases, driving them to kill their partners. Brad Steiger, in his book *The UFO Abductors* (1988., Berkley Books., N.Y.) describes Brian Scott's experiences as follows:

Brian Scott's first abduction reportedly occurred in the Arizona desert near Phoenix in 1971, and he claimed that another had just occurred on December 22, 1975, in Garden Grove, California. In between, Scott said, there were three other terrifying sessions with the aliens and repeated visits to his home by balls of light and a transparent being that called itself The Host.

Incredibly, Scott found that a friend of his was already inside the craft. The two of them were taken into a small room that began to fill with a fog or a mist. Then they were confronted by four or five 'very horrifying' creatures. Scott described them as having gray skin like that of a crocodile or a rhino, with a thicker patch of hide over the front torso. The beings were seven feet tall, according to Scott, and, had three fingers and a thumb kicked over to one side.

In his book, Steiger detailed a fascinating conversation between noted author and UFO investigator Timothy Green Beckley and Brian Scott.

Tim Beckley: What happened on the day when your wife was sent to the hospital?

Teleportation: From Star Trek To Tesla

Scott: She had been to work, pretty much handling everything that was going on around her. Then I got a call that she wasn't feeling very well. I brought her home, and after about fifteen minutes of sitting there talking with her, she was saying several things, none of which made any sense to me or to her. She said that she had been in the bathroom and suddenly felt hands all over her body. It was as if someone had broken into the house and molested her. When she calmed down and started making explanations to me about what the hell was wrong with her, it was as if, from her description, the guys I had seen aboard the craft in 1971 had visited her. This is odd, because she has never seen nor shown any interest in the sketches that I made of those entities.

Beckley: So this was an actual materialization – if you want to call it that – of the entities in the house?

Scott: I don't know what it was.

Beckley: But she was so upset that you decided to take her to the hospital?

Scott: Later that evening, it seemed as if she was okay. I was on the phone, and the baby was getting into everything so I couldn't carry on the conversation. I got up and went looking for my wife. I heard a bumping sound and a moan coming from the bathroom. My wife was on the floor, hyperventilating. I got her up and onto a chair in the living room. I was on my way to call her mother when she just fell flat on her face. I called the paramedics, and while they were on the way, she got up and fell down again. Then she began to become hysterical. It took four paramedics to hold her down. She was throwing people around as if they were tissue paper. Guys were thrown backward against the furniture. Finally they loaded her up in the ambulance. I came back in the house, and the baby was not in the playpen. I panicked, because I couldn't find our one-year-old baby who got out of a playpen!

Beckley: Who is The Host?

Scott: There is one entity that comes through that calls itself The Host, whatever that means. It speaks in what sounds like some kind of computerized language. The voice seems to come out of me, an inner voice that is not mine. The entity says that I am one with it. It says, 'I am; I am,' or 'You are one with me.' When I asked if it has a name, it will just come back and say, 'I am; I am.' The other night we heard some strange sounds coming from the bedroom, and I began to speak in a foreign language that we later found out was Greek. Where that came from, I don't know. I wrote in Greek backwards. On top of that, I was writing with my left hand, and I am right-handed. This voice was talking. We asked who it was, and the name Ashtar came out. Then it began to use the name Ashtar and speak to my wife. It told her things about her past that only she could know. This went on for a while, then it went on to say it would give her all the money in the world. It only wanted one thing in return – her soul.

Beckley commented that it sounded as though diabolical entities might be coming onto the scene, attracted by the extreme vibrations. He also observed that Ashtar, which has surfaced often in the strange world of UFOs, sounded very much like Ishtar, an ancient Babylonian goddess.

Tim Beckley later asked J.D. (an investigator associated with a civilian UFO investigations group who studied the Brian Scott incident) how he would differentiate between what may have originally been an abduction case and the various types of poltergeist phenomena that now seemed to prompt Scott's resultant trance state. Are they one and the same? Are they closely related mysteries? Or are they entirely different aspects of a more general phenomenon?

J.D. indicated that he was aware that there had been other cases such as Scott's. The manifestations of balls of light streaking through the homes of contactees and abductees apparently are more frequent than many investigators realize. J.D. mentioned that one voice, a horrible voice, came through and claimed to be Beelzebub, the Devil. J.D. was convinced that the entity was simply trying to frighten away the investigators.

Aside from the 3-fingered, 7 ft. tall "Crocodilian" creatures encountered by Brian Scott, there was another group involved in his abductions as well. According to Steiger: "The secondary group was composed of beings who were small, with frail bodies, milky white skin, large bald heads, thin lips, and enormous eyes... supposedly this group, perhaps from the sixth or seventh planets around the star Epsilon Bootes, have the power to veto actions planned by those beings of the secondary world, the reptile creatures."

In reference to the supposed mission these creatures had chosen for Brian Scott, Steiger states: "Scott was to design a transportation technology that would move matter through space. He was to master quantum displacement physics and begin to develop a mind transference machine to be used to unite all humans. Such a machine would help to develop a philosophy of cosmic brotherhood. The above tasks, of course, would seem impossible for a combination of Einstein and Superman, but they are typical of the type of grandiose mission(s) assigned to so many contactees and abductees."

The name Ashtar appears often in UFO contactee literature. One cannot help noting the ancient origin of the name Ishtar, Ashtar, Asta, described always as a god of evil and negativity in the Bible, but whose roots as a highly respected god go back to the beginnings of recorded history. Ashtar seems to belong more to the contactees than the abductees, but there are instances where those who claim to have been forcefully taken aboard UFOs describe an interaction with beings who represent themselves as emissaries of Ashtar's Grand Plan.

The Brian Scott case is just one of hundreds, maybe thousands of nonsensical interactions with nonhuman entities. As noted in Brad Steigers book, often UFO experiencers will be given some supposedly monumental task by the extraterrestrials, who solemnly tell their chosen emissary that the fate of the world depends on the successful completion of his mission. In the case of Brian Scott, his building of a super-scientific teleportation machine would help unite all of mankind with the aliens in a cosmic brotherhood. So far, Brian has yet to accomplish his goal – though it is one of the goals apparently embraced by many of the UFO episodes tied in with teleportation that we have presented in this book.

A Complex Picture of Unreality

The cases detailed in this chapter are just a few of hundreds of other comparable episodes that seem to defy rational explanation. UFOs, strange entities and their ability to mysteriously appear and disappear, or teleport, appear to have similarities with paranormal manifestations that display the same range of unusual tricks. It is obvious that these apparently different types of phenomena have more than just superficial similarities. They could be the same type of phenomenon, but wearing different faces as the moment demands.

This is not to suggest that these strange incidents are fantasies, or completely non-physical in nature. But as indicated in a number of reports, many seem to walk a fine line between "real" and "unreal." Evidence suggests that some experiences challenge tightly formulated definitions of reality, and may even point up the deficiencies in those definitions.

In the world of teleportation it appears that paranormal abilities, or the undiscovered powers of consciousness to effect reality, can transcend our three-dimensional world and move about with great impunity. This power can be seen in the antics of poltergeists and ghosts, as well as the more paranormal guises of the UFO phenomenon.

'Reality' is what we take to be true. What we take to be true is what we believe. What we believe is based upon our perceptions. What we perceive depends upon what we look for. what we look for depends upon what we think. What we think depends upon what we perceive. What we perceive determines what we believe. What we believe determines what we take to be true. What we take to be true is our reality.

Gary Zukav –*The Dancing Wu Li Masters: An Overview of the New Physics*

CHAPTER 10
URI GELLER AND THE SUPERMINDS

The notion that the mind, or consciousness, can manipulate matter through time and space has been demonstrated throughout history in poltergeist occurrences. A poltergeist is thought to be either the unconscious psychokinetic energy of a living individual, or a traditional haunting. Either way the manipulation of matter is controlled by an as yet unknown aspect of consciousness.

A poltergeist is considered to be an uncontrolled outburst of psychokinesis exhibiting an almost rudimentary intelligence. But what about controlled psychokinesis? Someone who is deliberately trying to use mind powers to manipulate matter. Is there any evidence that you can literally use the force of your "will" to move things around, or even make them disappear?

One such person who claims that he has the power to influence objects with his mind, and has the evidence to back up his claims, is Uri Geller. Uri Geller was born in Israel on December 20, 1946. His parents are of Hungarian and Austrian descent and he is distantly related on his mother's side to Sigmund Freud. At the age of four he had a mysterious encounter with a sphere of light while in a garden near his house. He has often wondered if this strange encounter was somehow responsible for his unusual abilities.

Geller first became aware of his powers when he was five. One day, during a meal, his spoon curled up in his hand and broke, although he had applied no physical pressure to it. His parents were shocked and Uri did not mention the incident to anyone else at that time. Later, he developed these powers in school with demonstrations to fellow pupils. His mother thought he had inherited them from Sigmund Freud.

When he was eleven, he went to live in Cyprus, where he remained until he was seventeen. He then returned to Israel, served as a paratrooper in the Israel army and fought in the Six-Day War of 1967 during which he was wounded in action. From 1968 to 1969 Uri worked as a model, he was photographed for many different advertisements.

In 1969 he began to demonstrate his powers of telepathy and psychokinesis to small audiences. By the end of 1971, however, his was a household name throughout Israel thanks to his numerous stage appearances. He was given a plug by then Prime Minister, Golda Meir. When asked on a national radio program what she predicted for the future of Israel, she replied, "Don't ask me – ask Uri Geller!"

In 1972, Uri left Israel for Europe, where he attracted widespread attention. In Germany, witnessed by reporters and photographers, he stopped a cable-car in mid-air using only the power of his mind. He then did the same to an escalator in a major department store. That same year he traveled to the United States at the invitation of Apollo 14 astronaut Captain

Edgar Mitchell, as well as inventor and author Andrija Puharich MD. Among the notable scientists he met were Professor Gerald Feinberg of Columbia University's physics Department, Ronald Hawke from the Lawrence Livermore National Laboratory, Ron Robertson of the Atomic Energy Commission and NASA's Dr Wernher von Braun, "Father of the Space Age," who testified that his own wedding ring bent in his hand without being touched at any time by Geller.

Geller toured the United States giving lectures and demonstrations, but comparatively few scientists were convinced that his powers were real and not the result of chicanery. He also performed on American television which resulted in hundreds of telephone calls from all across the country from families who reported cutlery bending in their home as they watched him on their TV sets.

In Britain Geller made several successful public television records of apparently paranormal bendings of cutlery, and many children came forward and claimed, sometimes even demonstrated, similar happenings. Mathematical physicist John Taylor, who was present in the studio, started a program of fieldwork and invited numbers of the children to his laboratory. He published accounts in a book entitled **Superminds**, and publicly affirmed that he believed paranormal metal-bending was a real effect.

As of the writing of this book, no one has yet to prove beyond a shadow of a doubt that Uri Gellers amazing abilities are not real. In fact, Geller has proved himself, at least in a financial way, by using his powers as a paid consultant to locate oil and valuable minerals. His success in this field has made him a very wealthy man. All of this from the humble origins of a spoon-bender.

The Geller Principle

The poltergeist phenomena has been extensively researched and studied over the years. Psychic researchers have sifted through firsthand accounts, judging them by the same criteria as those applied to historical, anthropological and forensic source material. Other psychic researchers have been fortunate enough to witness for themselves poltergeist phenomena, and have written some fascinating accounts as a result.

On this basis most scholars have concluded that strange, physical phenomena really does occur on a somewhat regular basis. The modern technical term for physical poltergeist phenomena is RSPK (recurrent spontaneous psychokinesis).

In a poltergeist case, objects have been observed to spontaneously fly about the house, apparently in a random way. Sometimes their paths of flight are unnaturally crooked, and often the objects are not seen to leave their normal positions; sometimes they just appear in the air, in full view of witnesses, and gently drop to the floor.

Teleportation: From Star Trek To Tesla

Some objects, when picked up immediately after their teleportation are found to be warm. Sometimes they arrive with spin, or angular momentum. Other times they appear in motion, their flight starting from a position different from their normal place. Sometimes the origin of the objects is unknown, as in the cases where showers of stones are reported.

Furniture can move spontaneously about the room, tip over, and even levitate and crash down again on the floor. Sometimes these movements are observed, sometimes they happen when no one is in the room.

A typical poltergeist case will last only for a period of weeks or months. The spontaneity of the movements has led to the opinion that the phenomena are caused by a ghost. However, the more usually accepted view is that one or more of the personalities involved, usually children, are unconsciously responsible for the phenomena, in that they are physically present when they occur.

Such personalities have been called epicenters (although this term is also used to describe the area of the house in which events most frequently happen). When the subject realizes that he or she is 'responsible,' and is now the center of attention, they will often add to the effects by normal physical means.

Rarer and stranger events, including many quasi-physical as opposed to physical phenomena, have been reported in poltergeist cases, but the above are the most usual and the most relevant to psychokinesis and teleportation.

John B. Hasted in his book *The Metal-Benders* (1981, Routledge & Kegan Paul) makes the observation that the particular type of poltergeist event he has observed most frequently is the traveling of an object from one location to another in an abnormal way. It might best be described as the disappearance of the object in its original position, and its re-appearance somewhere else.

In other words – teleportation. It is likely that this phenomenon is of physical similarity and relevance to metal-bending.

In November 1974, Hasted and his wife, Lynn, were paid a visit one Saturday afternoon by Uri Geller and two friends. Hasted had already met Uri on several occasions and had observed his metal-bending. But Lynn had never spoken to Uri and had never seen anything bend. She was strongly skeptical, and had never had the slightest interest in psychic phenomena.

Lynn served drinks in the lounge, and the guests commented on the carvings displayed on the piano and bookshelves. Lynn took Uri into the kitchen to get him an apple while the others stayed in the lounge. Lynn had started to tell Geller that she was entirely skeptical about metal-bending.

John Hasted was just entering the kitchen when he clearly saw a small object appear a few feet in the air and fall to the floor in front of the back door. Geller turned round to face it, and they saw that it was a small Japanese marine ivory statuette of an old peasant. This had been in its normal place on the bookshelf in the lounge. They were certain that the statuette had not been thrown; as it would have described a trajectory instead of dropping more or less straight downwards. If the statuette had been thrown by Geller, it would have bounced into the corner instead of dropping downwards. Moreover Geller had his back to its landing-place, and his hands were in front of him, with an apple in one of them.

While everyone was standing around looking at the statuette, a second object appeared in the air and dropped. This time everyone observed it, and it was clear that it had not been thrown. This object was the key of a Buhl clock which stands next to the statuette on the bookshelf in the lounge.

Hasted comments that if the statuette and the key had passed in normal parabolic arcs from the lounge bookcase straight to their destination in the kitchen, then the objects would have had to pass through a wall to get to the place at which they reappeared.

After Gellers visit, Hasted recorded over forty psychokinetic episodes in his house. One incident occurred on December 23, 1974 as the family was preparing a turkey for Christmas dinner.

The turkey was wrapped in a plastic bag and was resting on a tray on the bare white plastic table-top. Beside the turkey, on the tray and wrapped in another plastic bag fastened with wires, were the giblets, liver, etc. Suddenly a brown object appeared on the table in front of us, and I thought for a moment that it might be a leaf that had floated in through a window. But it was in fact a turkey liver, and we checked that one was no longer in the sealed plastic bag with the giblets. It resembled the other turkey liver, which we found to be safely in its own bag in the larder.

There was no smear of blood on the white table, such as the liver would have made if it had moved along the surface. There had been no sound, and there seemed to us no normal explanation. I did check with our butcher that it was actually a turkey liver.

This event was one of the most significant I had observed, since the liver in all reasonable certainty started from its situation inside the sealed plastic bag, and finished outside it. All three of us saw first of all an expanse of white table, and immediately afterwards a piece of liver on it. There were no holes in the plastic bag, although it was not vacuum-tight.

John Hasted continued to have teleportation experiences in the presence of Uri Geller. During July 1975 Hasted was exposed to a short but rapid sequence of teleportations while staying in the New Otani Hotel, Tokyo, in the room next to Uri Geller. About 10:30 PM one evening after a press conference during which there had been a miscalculation that upset Geller, his secretary Trina and photographer Shipi departed to send a Telex message. Hasted left Geller in his room, and unlocked the door to his room and went in. Within a few seconds he saw a small object fall to the floor, not from a great height, but within one foot of the drawn window curtains. It was a pair of nail-clippers.

When he took the clippers next door to Uri, he told Hasted that it belonged to him, and would normally be kept zipped-up in a leather case from which he showed that it was missing. While they were speaking, a glass tumbler dropped to the carpet behind them, in the center of the room.

Hasted took Geller back to his room to show him where the nail-clippers had fallen, but they got no further than opening the door when there was an explosion and crash. Broken glass was found all over the area by the door, and in the hotel corridor. One glass tumbler from the bathroom was now missing. Hotel guests in the corridor saw the flying glass but could offer no explanation.

Uri returned to his room and Hasted almost immediately saw the sudden appearance of his magnifying glass in the middle of the floor. By good luck it was reasonably well in his field of vision at the time, so that he was able to be certain that it did not just fall to the ground. Previously it had been on the desk, more than six feet away.

Based on his own observations, Hasted summarized some important features of the teleportation phenomenon. For obvious reasons, it is much more usual for the appearance than for the disappearance to be observed. Indeed, it would be very difficult to be certain that both things happen at exactly the same moment (say within 0.2 sec) without some instrumentation. On rare occasions the disappearance and reappearance locations are both within the field of vision.

The reappearance can take place either in the air or on a surface such as floor or table. Accounts exist of objects reappearing inside solids, particularly fruit. Teleportations into identified hens' eggs have also been reported.

Reappearing objects have often been observed to appear with their own angular momentum. I have seen objects spin rapidly as they fall to, or appear on, the floor. It is difficult to make generalizations about direction of rotation or orientation of the axis of spin.

In many poltergeist-induced flights, linear momentum is reported to be associated with an object at the moment of its reappearance. Detailed accounts by Maurice Gross, the Society for Psychical Research investigator of the Enfield poltergeist, mentions glass marbles which flew about the room. There was no question that these marbles were simply teleporting, and did not have flight trajectories; they could be seen in flight; but the problem of where their trajectories started was more difficult.

They seemed to start from the closed window, yet there were no marbles on the windowsill. The most likely alternative was that they had teleported to the window and appeared with linear momentum into the room. This feature might offer a clue to the 'dog-leg' flight-paths which are sometimes observed in poltergeist cases, and were seen at Enfield. In these paths there is a sudden change of direction in mid-flight. This might be interpreted as a teleportation in mid-flight, to a position almost identical with the point of disappearance, but with the appearance associated with a new linear momentum vector.

Sometimes the object after its reappearance is felt to be warmer than normal. Reports of the appearance of warm objects have appeared fairly frequently in the literature of poltergeist phenomena.

On rare occasions the disappearance of an object is observed several minutes or even hours before its reappearance. A disappearance is noticed, and at a later time the reappearance of the object is also observed. There is no question that the object was at the location of the reappearance for all of the interim period; this location remained exposed to the field of view of observers, and was usually an obvious one.

Very little has been observed which supports the hypothesis of gradual rather than sudden appearance. The gradual appearance and gradual disappearance of apparitions would seem to be a different phenomenon, at least as regards the long times of appearance. At the Stanford Research Institute, during Uri Geller's visit, an interesting video-tape was made of a wristwatch falling through the field of view onto a table.

Although it seems that the appearance took place above the field of view, the watch is seen to flicker as it descends; in consecutive frames the light reflected from the watch increases and decreases. One might be tempted to regard this as an 'oscillation in the intensity of the appearance,' but a more likely interpretation is that the presence of angular momentum causes the light reflected from the watch to vary periodically. A very interesting claim has been made by Dr Miyauchi that Masuaki Kiyota has materialized (or teleported) a full Coca-Cola bottle in stages; the bottom first, and then the top.

Information is slowly accumulating about teleportations of living creatures, including human beings. Hasted relates that he has never witnessed such events himself, but has received reports from various victims and from their families. Very little information is available about larger creatures, but Hasted once received a detailed report from Matthew Thompson, a poultry farmer in Dorset, which he summarized as follows: " I have within recent weeks had two separate instances of birds (caged chickens) disappearing and reappearing some hours later. I am talking about birds disappearing literally into thin air and being neither visible nor audible. Any possibility of them being removed by some other persons and then returned can be completely ruled out."

Andrija Puharich, who also spent much time studying Uri Geller recalled the time when Uri found himself unexpectedly teleported to Puharich's home.

Geller was with a woman named Maria Janis (Gary Cooper's daughter). He left the apartment they were in to go jogging. Within two minutes of leaving 68th and Park in New York, he somehow landed in Ossining which is 36 miles away. I was home alone. I heard this huge crash and thought it was an earthquake. I couldn't find the source of it at first, and then I heard this bleak voice, "Andrija! Andrija!" There he was crumpled up on the floor. He was intact and wasn't hurt, but the shock he experienced was considerable, and the event was never repeated. I've had a lot of that kind of stuff with Uri.

Geller remembers the incident very clearly: "I was jogging in Manhattan when the next thing I remembered was that I was thrown at something and I crashed through something on to a table then fell on the floor And I looked around and I was... you'll find it unbelievable, but I was in Ossining, a small town 36 miles out of Manhattan in the porch of Andrija Puharich.

I shouted his name 'Andrija! Andrija!' and it took him five minutes to find me. He went out of the house to look for footprints in the snow, but of course there were none."

One of the most unusual claims made by Puharich regarding the talents of Uri Geller involves Geller's alleged ability to teleport automobiles. Since Puharich's book, several other reports on auto teleportation have been filed.

Ray Stanford of Austin, Texas reported that after he used his car to pick up Geller at the airport, his automobile was apparently teleported. One of these involved an automobile accident. Testimony from witnesses is recorded in the traffic-court transcripts. They observed Stanford's car suddenly appear in front of them, "like a light that had been switched on." The distance of this teleportation was about fifty feet.

On the second occasion, the teleportation was even more dramatic. Driving with his wife along Interstate 10 in Texas, Ray suddenly noticed a silvery-metallic, blue glow around the car. Stuck in heavy traffic at the time, he actually mentioned to his wife he hoped "Uri's intelligences would teleport us away from here."

Then, according to his testimony he felt a strange sensation in his brain and the scene instantly changed. They had traveled thirty-seven miles in no time and using no gas. Later, as the car was not functioning well, it was hauled to a garage. The alternator and voltage regulator were completely burned out and all the wiring was completely charred.

One family of a young metal-bender told John Hasted that they would continually find their son in unnatural places, wedged in between wardrobe-top and ceiling, and so on. They would be running a hot bath for him, and suddenly a scream would announce his "transportation" from his bedroom into the overheated bath, for which he was totally unprepared. The affliction of this family lasted for several months, but eventually grew less serious.

One test subject named Nicholas Williams also claims to have been teleported out of a locked room. When his father pointed out that this left them with a problem of the key remaining on the inside, Nicholas teleported back again to unlock the door. He has described the experience as something like being in a blizzard.

Teleportation: From Star Trek To Tesla

The Superminds

Due to the extensive research done with Uri Geller and others like him, scientific communities in China, the United States and other countries are now identifying small groups of infants and children that display rare abilities such as purging HIV, advanced genius and psychic/telekinetic abilities and other extra-ordinary attributes.

Usually the phenomena are first noticed after a television appearance by Uri Geller. When the household first becomes aware of the bendings, it is often not known who is responsible. Nearly always it is one of the children, who finds that if he or she strokes a spoon between fingers and thumb it sometimes softens and bends.

These children can display some or all of these qualities and others not yet identified. In these children, fragments of DNA science identifies as 'junk DNA' and other portions of the DNA chain that science has yet to identify, are more organized and operational at birth than in the average populations, which gives these special children biological, mental and/or spiritual skills and abilities that appear advanced, compared to that of the norm.

One little understood attribute of Indigo advancement is that of perceptual expansion, an accelerated psycho-spiritual biological orientation and natural usage of sensory abilities that are beyond the range of the commonly known five senses. The phenomena of perceptual expansion due to genetic progress is presently evidenced and demonstrated in global culture through rapidly increasing occurrences and reports of unexplained events such as ESP, near death experiences, out of body experiences, angelic encounters, hauntings, inter-dimensional communications, teleportation, UFO abduction, lucid dreaming, etc.

Exactly how these superminds are producing psychokinetic energy to bend metal or teleport objects is not fully understood, but there seem to be two or three possible mechanisms, some combination of which may act at a given time.

❏ First there is evidence that a physical force may act on the metal and that this may be independent of the size of the region that it acts on.

❏ Second, groups of atoms may stop interacting with others and so be able to pass through from one location to another. This is the same phenomenon where objects pass through walls in poltergeist cases. An object may either be invisible as during teleportation or it can start to return, be visible, but not feel gravitational forces fully and gently float to the ground.

❏ Finally it returns to the real world. Beeping calculators have been teleported and the sound could be heard coming first from one location and then another, until it is at

Teleportation: From Star Trek To Tesla

last returned. Small radio transmitters have also been teleported by Chinese children. The possibility that there may be an as yet poorly understood interaction between particles and spacetime is discussed in theoretical physics text books.

It is acknowledged that curved spacetime can create particles. The reverse may occur under some as yet unknown conditions. What has not yet been investigated (as the full theory of quantum gravity is unknown) is the possible duality between systems of particles (and the information they contain) and gravitational waves / spacetime curvature.

Systems may be of greater interest because of the additional information that they contain, second quantization and so on. Quantum coherence allows pairs of particles to easily penetrate a barrier in a Josephson junction with a much enhanced probability and a similar effect occurs in the laser where the barrier to photon decay is lowered by the coherent radiation stream. Physicists are just starting to apply these ideas to coherent matter waves.

The final mechanism behind paranormal phenomena seems more common throughout recorded history than in the laboratory and involves more fundamental changes in reality such as substitution of parts of an object for another one or temporary duplication. It seems that some minds under some conditions can cause temporary construction of an object (such as a deceased person or duplication of their own body) partial dematerialization may also accompany this.

This could be a similar mechanism to the reality shifts observed when a psychic converts one object into another. It points to an abstract representation of objects in the world of physics and the mind being able to tap into this. This would also explain the claims of many psychics that they are not doing anything and that there must be some other external 'intelligence' involved.

There may be a natural hierarchy of abstract levels of representation in nature. Interestingly developing some of these skills seems to be on a fairly well mapped out path among meditators and Yoga masters. Such systems, if they exist, might be expected to behave more in accordance to our bureaucratic structures.

Uri Geller has his own opinion on the mysterious nature of teleportation: "I think it is some kind of automatic process with us, it's the sharpening of our inner powers, our mind, our brain, our spirit, our soul. I don't think we are going to build some kind of time machine. I think our minds will be able to teleport us. But I'm not talking around the corner, I'm talking 250,000 years from now."

CHAPTER 11
THE TESLA TELEPORTATION PROJECT

Whether the Philadelphia Experiment actually occurred as described has been highly debated. The concept, however, is valid, having been perhaps suggested or supported by Nikola Tesla's high energy field experiments and Einstein's unified field theory. In application both to the Philadelphia Experiment and the concept of natural teleportations, the attraction between molecules could be temporarily altered by a force field. It would, in effect, introduce matter into another dimension or, one might say, the etheric world. Such force fields comprise both the cause and effect of the transmutation and transference of matter.

The Philadelphia Experiment (Project Rainbow), at least in its early stages, probably began about 1932-33 in the Chicago area. It was at that time that a small team of scientists got together and started to investigate the subject and its possibilities. The team included John Hutchinson, Dean of the University of Chicago; Nikola Tesla, who became the project's first director, and Dr. Emil Kurtenauer, an Austrian, who had a Ph.D. in physics.

Although it was a private research operation, it was apparently funded by the Navy from the very beginning. It was soon moved to the east coast and wound up at the Institute for Advanced Studies at Princeton University, which had been formed in 1933. Einstein and a number of very interesting people became associated with the Institute, but one of the first to join them in 1933, was Dr. John Eric Von Neumann.

What Project Rainbow sought to do was to accomplish radar and visual invisibility for Navy ships. It is still unknown as to whether the original idea was to achieve invisibility to enemy radar or actual optical invisibility. Either way, it is commonly believed that the mechanism involved was the generation of an incredibly intense magnetic field around the ship, which would cause refraction or bending of light or radar waves around the ship, much like a mirage created by heated air over a road on a summer day.

In September 1940, Tesla had designed special coils and 500KW generators that were applied to an unmanned mine sweeper. Basically, a coil was wound on each half of the ship and driven by separate oscillators, synced with an adjustable phase angle to create a "Tesla scalar-type wave." This distorted the field matrices of matter encompassed within the field for "unusual effects."

Much to the surprise of everyone involved (with the possible exception of Tesla who had said that a high-energy, rotating magnetic field would produce optical invisibility), a successful transition into invisibility was accomplished. The next goal for the Navy was to put this equipment on a larger ship with a full crew.

Teleportation: From Star Trek To Tesla

However, Tesla became very concerned when he learned that the experiment would next include a live crew. He protested that the scientists and military planners had no true idea of what would happen to the personnel aboard the ship. He insisted that the Navy give Project Rainbow more time before the experiments would be conducted. The government told Tesla that due to the Japanese bombing of Pearl Harbor, it was now imperative to complete the project as quickly as possible. He was told to make the experiment work.

On a cold day in March, 1942, all the levers were pulled to place the battleship into invisibility – and nothing happened. Tesla had deliberately rigged the experiment to fail.

The apparent failure of the experiment permitted Tesla to say, "Well, it is obvious the experiment was unsuccessful." By this time, Tesla was a very sick man. He was still suffering from injuries sustained when he had been struck by a cab in 1937, and he desperately wanted to return to his home and pet pigeons. He left the project, and ten months later, on January 7, 1943, he was found dead in his hotel room in New York.

With Tesla now completely out of the project, the reins were handed over to Dr. John Von Neumann. Von Neumann felt that Tesla's concern for the human test subjects was unwarranted, and proceeded with the final testings with the U.S.S. Eldridge, DE (Destroyer Escort) 173, and a partial crew of human guinea pigs.

At 0900 hours, on July 22nd, 1943, the power to the generators was turned on, and the massive electromagnetic fields started to build up. A greenish fog was seen to slowly envelop the ship, concealing it from view. Then the fog itself is said to have disappeared, taking the Eldridge with it, leaving only undisturbed water where the ship had been anchored only moments before.

The scientists and Navy officials were ecstatic, the experiment had worked beyond their wildest dreams. The ship and crew were not only radar invisible but invisible to the eye as well. Fifteen minutes later the men were ordered to shut down the generators. The greenish fog slowly reappeared, and the Eldridge began to reappear as the fog subsided, but it was evident to all that something had gone wrong.

When boarded by personnel from shore, the crew above deck were found to be disoriented and nauseous. The Navy removed the crew, and shortly after obtained another. In the end, the Navy decided that they only wanted radar invisibility, and the equipment was altered.

On October 28, 1943, at 17:15, the final test on the Eldridge was performed. The electromagnetic field generators were turned on again, and the Eldridge became near-invisible; only a faint outline of the hull remained visible in the water. Everything was fine for the first few seconds, and then, in a blinding blue flash, the ship completely vanished.

Within seconds it reappeared miles away, in Norfolk, Virginia, and was seen for several minutes. The Eldridge then disappeared from Norfolk as mysteriously as it had arrived, and reappeared back in the Philadelphia Naval Yard. This time most of the sailors were violently sick. Some of the crew were simply "missing" never to return. Some went crazy, but, incredibly, five men were fused to the metal in the ship's structure.

The men that survived were never the same again. Those that lived were discharged as "mentally unfit" for duty, and sent to various mental institutions around the country.

Witnesses To Teleportation

The mistakes made during Project Rainbow were carefully swept under the rug of wartime secrecy. Nevertheless, there were a number of witnesses that were uncovered by William Moore for his book *The Philadelphia Experiment* (1973 Grosset & Dunlap). One witness was Victor Silverman who was an Engineer First Class, and one of three who were to pull the switches commencing the operation.

I was on that ship at the time of the experiment. There was enough radar equipment to fill a battleship, including an extra mast, rigged out like a Christmas tree. When the switches were thrown, the resulting whine was almost unbearable. Indistinct figures and other objects were present that seemed to not belong on the dock, if that was where I was. The green fog 'flashed off' and I wondered what in the world am I doing in Norfolk. I recognized Norfolk because I had been there before. Then suddenly the green fog returned and lifted and I found myself back in Philadelphia.

I passed the radar shack when I heard a party of three civilians commenting on the experiment and saying that, 'It was a great success.' When these three men left the ship, they were carrying a leather box about the size of a foot locker. For a long time afterwards, I was left wondering whether or not I hadn't lost my mind for those brief moments. I still don't know exactly what happened to me or to the other men who were there.

There is also a story uncovered by researcher Tony Wells in which five English merchant seamen were waiting in Norfolk, VA for berths on Liberty Ships bound for England. The Englishmen reportedly saw a strange sea-level cloud suddenly forming in the harbor, then dissipate, leaving a destroyer escort. The ship stayed for a few minutes, then disappeared under the same type of greenish fog. Within minutes, the whole area was cleared by Naval Security and the Shore Patrol.

Frederick Tracy, another WWII veteran, served aboard the USS Antietam (CV-36). Tracy heard about the experiment from a shipmate, D.J. Myers, when their ship was in drydock in Philadelphia.

"See that dock over there? That's where a ship once disappeared." Myers proceeded to tell Tracy of the Experiment, to which Tracy responded, "I don't believe you." Myers rebutted, "I didn't think you would."

The next year, the Antietam went to the Philadelphia Navy Yard for degaussing, when a rumor started about another Philadelphia Experiment. The Commanding Officer of the Antietam, Captain Teague, called the crew to quarters and read a memo from the Secretary of the Navy.

According to Tracy's recollection, the memo cited an event that happened on October 28, 1943, but concerned a patrol craft that vanished during a degaussing experiment. The patrol craft (PC) appeared in Norfolk, then disappeared only to reappear in Philadelphia. Despite the fact that the whole crew joked and half-heartedly believed the memo, the CO stressed that "to mention or talk about this incident outside the confines of the vessel would constitute an act of treason."

But an interesting twist occurred with the names of the ships; the Navy ship became the Coast Guard ship Eldridge, while the Coast Guard ship became the Navy ship Yarmawa. As far as the crew is concerned, Tracy states, "The only thing I ever heard about the crew of the Eldridge was that they were kept at Bethesda Naval Hospital out of communication with everybody for the remainder of the war."

In *The Philadelphia Experiment*, an electronic construction specialist by the name of Patrick Macey related a strange story to William Moore about a co-worker by the name of "Jim." While working in Los Angeles in 1977, Macey and Jim were talking about how much information the government was covering up about UFOs when Jim told of a strange experiment he had seen on a film:

I had an unusual experience during WWII, while I was in the Navy. Not with UFOs, but something pretty mysterious. I was a guard for classified audio-visual material, and in late 1945, I was in a position, while on duty in Washington, to see part of a film viewed by a lot of Navy brass, pertaining to an experiment done at sea. I remember only part of the film, as my security duties did not permit me to sit and look at it like the others. I didn't know what was going on, since it was without commentary. I do remember that it concerned three ships. When they rolled the film, it showed two other ships feeding some sort of energy into the central ship. I thought it was sound waves, but I didn't know, since I, naturally, wasn't in on the briefing.

After a time the central ship, a destroyer, disappeared slowly into a transparent fog until all that could be seen was an imprint of that ship in the water. Then, when the field, or whatever it was, was turned off, the ship reappeared slowly out of thin fog. Apparently that was the end of the film, and I overheard some of the men in the room discussing it. Some thought that the field had been left on too long and that had caused he problems that some of the crew members were having. Somebody mentioned an incident where one of the crewmen apparently disappeared while drinking in a bar. Somebody else commented that the crew were still not in their right minds, and may never be. There were also references to some of the crew having vanished permanently. At that point the conversation was carried on outside of my hearing.

Project Rainbow Continues Undercover

Allegedly, Project Rainbow resurfaced in 1947 or 1948, and was funded to continue research into the phenomena encountered on the USS Eldridge. This project was concerned with the electromagnetic technology. Dr. John Von Neumann and his research team, loaned to the Manhattan Project during the Second World War, were called back and put to work on a new agenda. It was similar to Rainbow but had a different goal. They were to find out how to protect humans within an electromagnetic field so vessels and crews could be transported through space and time without harm.

By the early 1950's, Project Rainbow and a weather modification program were included under the same funding and referred to as the *Phoenix Project* . Dr. Von Neumann was placed in charge of Phoenix because of his earlier work in advanced concepts of space and time. He originated and built the first vacuum tube computer at Princeton University, where he served as the head of the Institute for Advanced Study.

Von Neumann quickly learned that he was going to have to study metaphysics, to understand the metaphysical side of man. The Rainbow experiment had disassembled the physical and biological structure of human beings. Crew members had melded with bulkheads and had physically changed beyond recognition. Those who had survived were quite mad or died later from madness and spontaneous combustion.

Project engineers and scientists spent nearly ten years working out why human beings had troubles with electromagnetic fields that lofted them through different times and spaces. It now appears they discovered that humans are born with what is known as a "time reference" point. At conception, an energy being (human) is attached to a time line and must begin life manifest as flesh from that point. To understand this, it is necessary to view the "energy being" or soul as completely different from the physical body.

Our references as both a physical and metaphysical being appears to have origins in the time reference residing within the electromagnetic background of Earth. This time reference is the basic orientation point to the way the universe operates. Time, in the normal context, appears always to flow in one direction, forward, at least to our limited senses.

But the Rainbow technology apparently creates an alternate reality having its existence entirely within the quantum field, literally transferring material objects out of our "normal" time and space. This accounts for the light-invisibility of the USS Eldridge and her crew.

Teleportation: From Star Trek To Tesla

The alternate reality created by the shift has no time references because it is not part of the normal forward flow of time. For the person who was experiencing the phenomenon uninstructed, it would be like having an intense and enduring nightmare where nothing makes any sense.

So Phoenix engineers were faced with solving the problem of letting human test subjects get into and out of the quantum field without harm by somehow connecting them to the time reference they could recognize as the planet Earth. This meant that when they were in the alternate reality, they had to be equipped mentally with something that would give them a "real time" reference.

Engineers solved this by feeding in all the required natural background information of the Earth to convince test subjects of a continuous stream real Earth time reference so they would not experience trans-dimensional disorders.

Dr. Von Neumann was the ideal director for the Phoenix Projects. He knew computers would have to be used if they were going to calculate the time references of specific people and replicate those references while they were in the quantum field, otherwise the test subjects would be experiencing essentially "no reality" or a continuous nightmare reality at best.

The computer had to be programmed to generate an electromagnetic background with which the test subjects could synchronize. If not done, the soul and the physical body time and space reference points would be out of sync, resulting in dissolution and insanity or inability to return to Earth real time.

Because they were dealing with two separate and distinct entities – the spiritual human and the physical human – the time reference would be required to lock in the spirit and the electromagnetic background would be required to lock in the body. The technology, began in 1948, and was apparently fully developed by 1967.

When the project's first phase was complete, a report was submitted to the Congressional committee from which funds had been appropriated. Congressional members were told that the consciousness of man could definitely be affected by electromagnetism and, additionally, that it would be possible to develop electronic hardware and software that could literally change the way a person thinks about anything and everything.

The committee, fearing they would be first on the list of new test subjects, refused to refinance and Project Phoenix was disbanded in 1969. But the scientists and engineers at Brookhaven had spent too much time, effort and money on Phoenix to just scrap it and walk away. The technology was complete, and the engineers were looking for a mission. What they needed was funding from a secret agency to continue with the mind control experiments. The military seemed the most logical source.

When told a device had been developed that would alter the way people, and particularly soldiers, would think and act, the military was at first skeptical. However, after being shown the amazing capabilities of Project Phoenix, they agreed to provide the funds to further develop the project. Such a device could not only change the outcome of a battle already begun, it might be used to convince entire populations that war is not only unnecessary, but completely impossible. Or that total war was the only solution to a political crisis.

Exploration into telepathy, teleportation, transportation, levitation and tentative excursions backward and forward in time has continued almost without cease from 1943 until the present time. The results of the experiments aboard the USS Eldridge, disastrous as they might have been at the time, provided fuel and data for a series of programs and black projects that now supposedly consumes a great portion of America's defense budgets.

Scientists and engineers learned how to alter and direct weather, how to create storms or to vanquish natural storms by tapping into the planet's orgone energy and turning it on or off at will.

One facet of the Phoenix Project was intense research into various areas of paranormal activity, particularly telepathy. The military was well aware that the Soviets were involved in paranormal research, and they knew that if sensitive psychics could be found who possessed skills or powers strong enough to nullify Soviet psychics or even overpower them, America might have an incredible super-weapon with which they could defeat any foe, military or civilian.

By 1970 the Brookhaven group as well as several elite universities worldwide were deeply involved in paranormal research. Some of these splinter projects involved use of drugs and powerful hallucinogens. Some experiments were conducted using electromagnetic fields. A few purists tapped only the mind-power of their test subjects. These fascinating experiments convinced scientists that feats such as teleportation could be accomplished using electronic technology, and even more incredible, the simple power of the human mind.

Those who wish complete details on the Project Rainbow experiments should read the *Montauk Project* series edited by Peter Moon. The bulk of the material is derived from the research and personal experiences of Preston Nichols and the lone survivor of the Philadelphia Experiment, Al Bielek. This material is updated frequently and is highly recommended for those interested in this fascinating story.

Separating Fact From Fiction

The Philadelphia Experiment story by now has been added onto so much over the years that it is now impossible to separate fact from fiction. For instance, a caller who phoned in to KTKK (K-Talk) radio in Salt Lake City, in August of 1991, made some rather startling claims concerning high-tech Philadelphia-type experiments that had allegedly been conducted at the Top Secret Labs in or below the Nevada Military Complex.

The caller referred to Sean Morton who had allegedly visited Area 51 and talked to a man who said that most of the technology depicted in the first Star Trek series of the 1960's is now in the possession of the Secret Government. Morton also was told that, possibly prior to the Philadelphia Experiments, top scientists who were tied-in with the secret government succeeded in teleporting a china plate and that they were surprised that the object materialized in reverse image, including the lettering.

After some time they allegedly worked out the bugs and attempted to teleport a human being. The first human being to be teleported (apparently after several successful experiments with animals, etc.) was the man known among some New Age groups – and several government officials as Maitreya.

This man, according to Morton, was eventually teleported and made to materialize in front of a stunned African tribe as an experiment to determine how successful the Secret Government could be in future attempts to pass-off Maitreya, or others like him, as gods or divine entities in order to manipulate the religious passions of certain groups.

This story, which comes from Branton's online book *The Dulce Files*, sounds familiar to anyone who has studied UFO phenomenon. UFO literature is full of accounts of people and objects passing through solid objects and materials such as walls and windows. Additionally, there are a number of reports of highly structured UFOs seen entering the ground by simply passing through the surface.

This phenomena is hard to explain without a "frequency" theory of matter organization, as the simplest and most direct way to pass solid objects through each other is by adjusting the frequencies of atomic components so that their parts do not interact with other parts. Since 99% of all normal matter is "empty space," there is plenty of room for properly adjusted vibrating structures like atoms and electrons to sidestep, or completely pass through each other.

Density changes have been offered as a mechanism for such passing through walls, but the problems involved in density variability are great. To lower the mass density in any sector of spacetime, matter or energy must be removed from the frame of reference, or it must be rarified by expansion. We don't seem to see evidence of such momentum transfer, as the wall isn't reported to move aside or expand, but rather, a strange penetrating sensation is reported, as if individual atoms are slipping past each other.

There is also a very real possibility that the body "tunnels" through the wall by quantum mechanical means. Tunneling happens because any object is spread out in space in the form of a complex wave field, and there is always some probability that it may interact near the far edges of the wave field, and thus be said to exist there. The effect is nowadays well known, forming the basis of several important high-tech devices such as the tunneling electron microscope and the tunnel diode.

Teleportation: From Star Trek To Tesla

But for a large object to tunnel, however, requires that all its components be rendered substantially coherent, so that all of it will be found on the opposite side of a barrier at the same time. This is extremely difficult, because large objects have very complex superposed wave forms, which would tend to leave most of it out of phase, somewhere else besides the hoped for destination.

Right now, only single or small groups of electrons are routinely used for tunneling purposes, but now that we have the ability to get clusters of atoms to act as single units when cooled to near absolute zero (Bose-Einstein condensates), it is at least possible that someday whole molecules might be induced to tunnel, although doing so at room temperature may be forever impossible with presently understood physics.

Getting large groups of organic molecules in masses the size of a human body to do so at room temperature is definitely not within the range of current theory. Therefore, if UFOs, or for that matter, poltergeists and people like Uri Geller, use tunneling processes to do their passing through walls tricks, then there is probably a new physics of getting large structures to act coherently, allowing them to avoid interfering with themselves and the objects they pass through.

It is equally likely that teleportation is not a tunneling process in the presently understood sense, since tunneling is almost always associated with strong voltage differences, which do not seem to be in evidence in every poltergeist, or UFO related event of this nature.

Teleportation by this effect could be an explanation for the accidental results of the first Philadelphia Experiments. The process was poorly understood and resulted in accidents and the unfortunate injuries and deaths of those who were forced to participate. Even today, civilian sector research on tunneling by electrically neutral large objects at room temperature is still inadequate and poorly funded.

The main reason for the lack of a satisfactory explanation is that the phenomenon cannot be understood given modern understanding of the quantum mechanical process. It is more likely that a whole new, completely different set of physical processes must be discovered in order to gain a more complete knowledge of the mysteries of teleportation.

CHAPTER 12

PRACTICAL TELEPORTATION

In previous chapters we have examined the many different varieties of possible teleportation phenomenon. Modern scientific theory says that some form of teleportation could be possible, but the cold harsh realities of making it actually work is far beyond our current understanding of spacetime and how to influence it.

However, if we are to believe such stories as the Philadelphia Experiment and others, then teleportation may not be as difficult to accomplish as we think. Evidence in the form of UFO experiences where teleportation was possibly utilized, to the incredible abilities of such super minds as Uri Geller, indicate that teleportation is a reality.

In order to accomplish teleportation we must look at some of the ways that have been used in the past. The Philadelphia Experiment is probably the most well known in the utilization of electronics to achieve teleportation. However, it is unlikely that most home hobbyists could ever hope to even come close to the Tesla-based science that was used on the USS Eldridge in 1943.

John Hutchison, an electromagnetics pioneer in British Columbia, Canada, thinks that he may have discovered at least one of the secrets behind poltergeists and spontaneous psychokinetic activity. A secret that could also make a form of teleportation available to anyone with a little electronic ability.

After years of research and experimentation, Hutchison believes that poltergeist activity is electromagnetic in nature. The Hutchison Effect is a collection of phenomena which were discovered accidentally in 1979 during attempts to study the longitudinal waves of Tesla. In other words, the Hutchison Effect is not simply a singular effect. It is many. The Hutchison Effect occurs as the result of radio wave interferences in a zone of spatial volume encompassed by high voltage sources, usually a Van de Graff generator, and Tesla coils.

What he has done is cram into a single room a variety of devices which emit electromagnetic fields (such as Tesla coils, van de Graaff generators, RF transmitters, signal generators, etc.). He found that after the machines had been running for a while, effects began to occur that were identical to what have come to be regarded as poltergeist phenomena.

Objects of any material levitated into the air and hovered, or moved about and then fell; fires started in unlikely places around the building; a mirror smashed at a distance of 80 feet away; metal distorted and broke; water spontaneously swirled in containers; lights appeared in the air and then vanished; objects would teleport, or disappear completely; metal became white-hot but did not burn any surrounding materials.

Teleportation: From Star Trek To Tesla

Everything that psychical researchers have been documenting for decades as poltergeist activity, eventually turned up in the laboratory where John Hutchison's device operated. Although it was made up of different parts, it operated as a single machine, and phenomena occurred in the same unpredictable way as reported poltergeists: you could be there for days and nothing would happen, then suddenly coins would flip and fly, things would appear and disappear, water would swirl and a transformer would blow.

On one video recording a 19-pound bronze cylinder is seen to rise majestically into the air, at a distance of 80 feet from the center of the device, but, incredibly, Hutchison says: "The source power was 110 volts AC. One side of the AC line had a power factor capacitor (60 cycles, 250 volts) and a 100-amp current limiter."

On another occasion, when Hutchison's layout of apparatus and equipment was reproduced by an electrical engineering company interested in this device, he explained: "All components are powered from a single 15-amp, 110-volt, 60-Hz supply."

Mark Solis, a researcher who maintains a website devoted to the experiments of John Hutchison, notes that the fusion of dissimilar materials, which is exceedingly remarkable, indicates clearly that the Hutchison Effect has a powerful influence on Van der Waals forces. In a striking and baffling contradiction, dissimilar substances can simply "come together," yet the individual substances do not dissociate.

A block of wood can simply "sink into" a metal bar, yet neither the metal bar nor the block of wood come apart. Also, there is no evidence of displacement, such as would occur if, for example, one were to sink a stone into a bowl of water. This same effect was observed during the Philadelphia Experiment.

The spontaneous fracturing of metals, as occurs with the Hutchison Effect, is unique for two reasons: (1) there is no evidence of an "external force" causing the fracturing, and (2) the method by which the metal separates involves a sliding motion in a sideways direction, horizontally. The metal simply comes apart.

Some temporary changes in the crystalline structure and physical properties of metals are somewhat reminiscent of the spoon bending of Uri Geller, except that there is no one near the metal samples when the changes take place. One video shows a spoon flapping up and down like a limp rag in a stiff breeze.

The radio wave interferences involved in producing these effects are produced from as many as four and five different radio sources, all operating at low power. However, the zone in which the interferences take place is stressed by hundreds of kilovolts.

It is surmised by some researchers that what Hutchison has done is tap into the Zero Point Energy. This energy gets its name from the fact that it is evidenced by oscillations at zero degrees Kelvin, where supposedly all activity in an atom ceases. The energy is associated with the spontaneous emission and annihilation of electrons and positrons coming from what is called the quantum vacuum.

Teleportation: From Star Trek To Tesla

The density of the energy contained in the quantum vacuum is estimated by some at ten to the thirteenth Joules per cubic centimeter, which is reportedly sufficient to boil off the Earth's oceans in a matter of moments. Given access to such energies, it is a small wonder that the Hutchison Effect produces such bizarre phenomena.

For example, part of the Hutchison effect literally rips half-inch- square steel bars apart and actually shreds the shattered ends, all at low power and at a distance. Tremendous energies come from somewhere, and in his experiments with the disruption of metal masses in the laboratory, Hutchison has developed his own ideas. He wonders if somehow the fabric of space-time is actually breached. As he puts it:

The idea is to excite the surface skin of the masses and their atoms to create an unstable space-time situation. This might allow the fields from the Tesla coils and RF-generation equipment to lock up in a local space-time situation. My thought is that now a small amount of energy is released from the vast reservoir in space-time at the sub-atomic level to create a disruptive or movement effect.

Physicists and electrical engineers should now reconsider the nature of severely modulated electromagnetic fields, for there are evidently previously unrealized potentials. The energies involved in the Hutchison Effect are clearly the same ones at work during some types of poltergeist activity. The missing factor seems to be the proximity of a consciousness that can direct these electromagnetic energy potentials. In other words: the Hutchison Effect, as well as other similar phenomenon, requires the catalyst of a conscious (or subconscious) mind to achieve the desired effect.

At the present time, the phenomena is difficult to reproduce with any regularity. The focus for the future, then, is first to increase the frequency of occurrence of the effects, then to achieve some degree of precision in their control. The work is continuing at this time.

Since we don't have access to such incredible technologies as imagined by such modern day geniuses as Nikola Tesla; we must instead look at how teleportation can be accomplished by using the amazing powers of the one thing that we all have in common, the human mind. This seems to be the one consistency in spontaneous cases of teleportation – the presence of people who have consciously, or unconsciously, made use of a hidden faculty within us all to physically transcend spacetime. The question is: "How can we learn to use this ability whenever we want?"

For centuries, shamans, priests, magicians, and others who were more connected to the non-physical world than the rest of us, have said that our world, our reality, is merely an illusion. An illusion of coherence in a universe made up of swirling energies and potentials. This illusion is given order by the observation of our consciousness. We literally create and control our universe and its laws with the power of our minds.

First Steps To Teleportation – Learning to Use PK

If teleportation can be achieved with the power of our minds, then some training to unleash its hidden potentials is desired to those who may not be "naturals" at using their mind powers. One aspect that seems to have a direct connection to teleportation is the power of psychokinesis (PK). To learn teleportation, one must first learn to control PK.

PK is essentially the ability to move an object on the physical plane using only psychic power. While some people think that it is an occult practice, this is not strictly so. We are all born with this skill. It is inherent, like walking, talking, breathing. We simply don't realize that we have the ability to use and control this power.

A common theory is that PK works by energy fields (magnetic or electric) or by "waves" of psychic energy which are actually dense enough to push/repel an object or draw it inward. Most people's only encounter with PK is accidental, something mysteriously falls over or objects fly around a room. As we have noted earlier, this is a phenomena often mistaken for a poltergeist when it may actually be a person with spontaneous psychokinetic powers. Anyone can harness their power and use it with the proper devotion.

❑ Meditate daily for half an hour, fifteen minutes if you're schedule is too busy.

❑ Attempt PK at least once a day, twice if possible. Give yourself a good 30-60 minutes.

❑ Focus on one method for at least a week, if it shows no results, switch methods.

❑ Be at ease; instead of taking it too seriously think of it as an experiment, a game. If you try too hard you'll just end up frustrating yourself and you'll get nowhere.

❑ Don't give up. Don't tell yourself you can't do it, because you can if you believe you can.

There are certain things you need to work on before you can get anywhere with these first steps in teleportation. Here are some of the obstacles that must be overcome and the exercises that must be practiced.

Doubt: PK is real. People have done it, people do it all the time. It is a gift all humans hold. Get that through your head before going any further. Doubt is like a wall between you and PK. So climb over the wall and leave it far behind you. You do not need doubt in your search for truth, especially not when it comes to the powers of the mind.

Teleportation: From Star Trek To Tesla

Logic: We live in a practical world, it's only natural to try to reason things. Don't bother. Set practicality aside, and when PK and teleportation is achieved, it then becomes clear how silly the concept of logic is altogether. It will all make sense when the goal is accomplished.

Purity of the Heart: Why do you want to teleport? Your reasons must be positive and your intent must be pure. If you wish to learn it to harm people, to profit off a strange talent, or just to impress your friends, forget about it. Learn it to exercise the mind, learn it to challenge your reality. Don't try it for some kind of cheap glory. If that's your intent, you'll get nowhere.

Meditation: You must meditate. Consider this a requirement. Take it one step at a time. Meditation is key in order to prepare the mind for psychokinesis. The most important thing to understand is that it is possible. PK is part of physical reality, just as telepathy is. The next most important knowledge related to achieving success with this, or any other human ability, is realizing that if one person can do it, any normal person can do it.

There is a prerequisite to achieving success in this endeavor, and that is having the ability to relax completely and focus your attention without distraction. In itself, concentration is one of the easiest activities one can possibly perform. It is simply being aware of one thing without distraction.

In actual practice it can be nearly impossible to maintain concentration for any length of time due to the subconscious concerns of day to day living. If you are going to practice the techniques described below, it would be to your advantage to perform relaxation exercises prior to beginning. After you have achieved success a few times the necessary focus can be reached quite easily.

PK could be the manipulation of a sort of human "magnetic field" around the body which can be concentrated in a specific area or distance. By concentrating this magnetic-like energy, objects can be moved or even teleported.

Using the analogy of magnetism is a good way to focus your attention. You need to imagine a magnetic-like field around your hand, and push your hand (or finger tips) into the "magnetic field" of the object you want to move. When one magnetic field interacts with another it will either push or pull against the other depending on the two magnetic field polarities. Opposite fields attract, similar fields repel. By concentrating on the belief that the field around my finger tips was very strong, I imagine the magnetic field moving the object.

Psychic phenomena happens at a non-verbal level of awareness. Your conscious thoughts only help focus your attention. They don't do any of the work involved. If your attention is on the thoughts you think you will not be able to maintain any consistent results. Your thoughts direct your intent to the subconscious, which then does what you honestly expect it to do. It's not what you think, but what you expect. Once you have achieved certainty that you have the ability to do this, it is far easier to expect the results you want.

117

Teleportation: From Star Trek To Tesla

There is another preliminary exercise you can do in the beginning to help focus your attention. If you hold your hands together palm to palm (as in prayer) but space them ever so slightly apart so they nearly touch, but don't, you can feel the heat given off by each hand. It is a very subtle sensation but you can indeed feel it. If you then "imagine" that what you are feeling is actually the magnetic-like field emanating from each hand, by moving your hands a bit closer together you can feel the fields pushing against each other. When you move your hands apart slightly, you can feel the pull of the fields.

It is believed that these fields actually exist, and that the subconscious is aware of them. By concentrating your attention on your perception, you are simply tuning into subconscious awareness. This isn't all that unusual, since right now your subconscious is aware of whether you are breathing in or out, and if you are leaning forward or to one side slightly, etc. As soon as you think about it you become consciously aware of subconscious awareness that was always just below the surface.

Moving a Candle Flame

First, meditate for half an hour. Light a candle in a safe place. A white candle on a white/light colored table is best.

Get comfortable and breathe deeply. Stare at the candle's flame. Do not think about anything, clear the mind but keep the flame in front of you. Soon you should see only the flame. Not the table, not the candle, not the room and objects surrounding you, just the flame.

Now imagine the flame is stretching upward, growing taller, brighter. You can put your hand above it and imagine you are drawing the flame upward. Imagine the flame shrinks, becomes smaller and shorter. Imagine it flickers and dances. Imagine it bends. Practice these things until you feel you are "one" with the flame, and it is doing as you desire it to.

Keep practicing this, once you are comfortable with your ability to do this, try putting the candle out with your mind. This exercise can also be done with attempting to move the rising smoke from a stick of incense.

Moving a Small Object

If you're eventually going to teleport yourself from place to place, you will have to start with something small. Choose a lightweight object, preferably made of a light metal, such as a cheap ring. Clear your mind completely. You should have no distractions whatsoever, and try not to let erroneous thoughts into your mind, or you will lose your focus. Concentration is vital.

Teleportation: From Star Trek To Tesla

Build a "tunnel" between you and the object. Visualize this tunnel between yourself and the object. You only see the object. Everything else is outside of the tunnel. Now, imagine your mind's hands coming out and pulling the object in. Once you feel the pull, build up a strong vibrational field and imagine the object vanishing a reappearing somewhere else. Don't expect this to work the first time you try it. Keep practicing.

Bending Forks - Spoons - Keys

Find a utensil of your choice and hold it in your hand/hands, however you are comfortable. Sit quietly, breathe comfortably, relax. Now, empty your mind of all extra chattering and thoughts. Remember to stay focused.

With your eyes closed, slowly rub your fingers tips over the surface of the object. Feel, don't think about it. Experience what the surface feels like. Get into the flow of molecules, atoms, energy. This may take a few attempts, but you will begin to actually "feel" the energy as it flows through the object.

At that very moment when you feel that you and the object are a blend of energy, just bend it. If you've done it correctly, it will bend. Remember never apply force. You aren't there to physically force it to bend. That's not the point of the exercise.

The Compass Exercise

Place a compass flat on a stable surface and place one or both hands about one or two inches above the compass. Don't think about anything. Just relax holding your hand above the compass and have a knowing resolve of what you are there to do. Then just allow the neural network in your brain to do what it knows how to do.

Be mindful, however, of why you are there and what you are doing. Don't just think about it. You are not suppose to be engaging the cognitive part of your brain. You may or may not feel the energy surge through your arms and fingers. That's all there is to this exercise. Simple, easy and to the point. Your goal and only goal at this juncture is to get the needle to move. A word of caution. Don't use an expensive compass, as usually this exercise will ruin a compass forever. This energy will alter certain structures.

Cork and Water Exercise

This method is very simple to put together. All you need is a bowl of water, a cork, and a small paper clip. You can either glue the paper clip to the top of the cork, if you like, as all you are trying to do is add a little weight to the cork so the cork doesn't float around on its own by room current or room air flow. The other method is to make a groove in the top of the cork. Uncoil the paper clip and let it rest in the groove. Either way, both work just fine.

Place your hands about one or two inches above the cork. Release all preconceived ideas as to what should be happening next. Just let the energy flow. Feel it move through your arms and out through your finger tips. In a really good session you can have that cork sailing all over the bowl.

Swinging On a String

Meditate for half an hour. Cut a piece of thread about 10 inches long. Tie a wide object to it, like a needle, a toothpick, or a short piece of a wooden rod. Hang the string so that you can sit comfortably in front of the hanging object.

Sit far enough away from the string so that you know you aren't making the object spin with your breathing. Let the object come to a complete stop, not turning at all in midair. Now, take a deep breath.

Focus all your attention on one side of the object, right or left, it doesn't matter which. Imagine the object slowly turning as if your mind was really pushing against that side of it. You can use your fingers to "point" your energy, but do not touch the object.

Once it starts to turn, and who knows how long it will take you to get this far, imagine that your energy pushes it faster so it speeds up. It should turn with greater momentum. Practice this. Once you get the hang of it, keep using this method to get the object to swing instead of just spin.

Why Isn't This Working For Me?

Remember, there are laws governing these principles. We may not understand or even be cognizant of them, but they are there and they do work. One of the reasons it is so difficult to get things to lift or slide has to do with friction, resistance, etc. You are trying to get a bowl to slide across the counter. Ask yourself about the dynamics involved here. Do you realize just how much energy it takes to perform that event?

It's a lot harder to move something than to bend a spoon or fork. Why? Because you are dealing with things like horizontal distances between the object's center of mass, the point of contact to the surfaces. It's actually easier to levitate or even teleport something then to drag it across a table.

Remember, that you want to start with easier things first, then if you find you have an ability, graduate to more difficult projects. Don't be afraid to experiment, just realize that teleportation is possible and stick with your goal.

Teleportation: From Star Trek To Tesla

Dream Your Way To Teleportation

As with rediscovering your potential for psychokinesis and its connection to teleportation, learning to astral project can also open up hidden abilities to the potentials of teleportation.

It seems strange, somehow, that one of the greatest clues to the nature of the universe should be sleep. It is believed that the mind is completely independent of the brain. A physicist is likely to wonder where the Mind gets its energy. It is certain that mental activity consumes a lot of energy, indeed, can consume far more than vigorous physical activity.

The mind is composed of a completely different kind of matter than the body. We know this because the mind is invisible to physical eyes and cannot be photographed in any reproducible way. When we open up the head in a living subject to perform a brain operation, there is no sign of something different, i.e. the mind.

The first law of mind is that it is not influenced by the electromagnetic force. However, the mind can effect the electromagnetic force. A common aspect of out-of-body experiences (OOBEs), is that a person in that state can freely move through solid objects, such as people, or doors, walls, ceilings, floors, whatever. We know that the tangibility and visibility of objects is entirely due to the interaction of their component atoms with the electromagnetic field. The mind has inherent ability to effortlessly warp spacetime.

The practice of astral projection can help teach you how to warp spacetime when in the awake, physical state and realize teleportation. This is an ability not to be taken lightly. It will require a lot of practice, and possibly years of self-discipline before anything can happen. But, the final results are certainly worth the effort.

Astral projection is a popular area of occult literature; for traveling to other worlds and places while the physical body sleeps is an exciting notion. Astral projection is not considered dangerous. It is as safe as sleeping. Most dreams are probably unconscious astral projections, and so far, we have all returned safely from our dreams.

Although there has been quite a bit written on the subject, astral projection is difficult for many people. The main difficulty is the tendency to forget dream consciousness upon awakening. Accordingly, the successful practice of astral projection requires work and practice.

Modern psychology discounts the idea of actual OOBE (that the spirit temporarily vacates the physical body). However, the idea is very ancient and has been reported throughout recorded history.

The Tibetans have an entire system of yoga (dream yoga) based upon astral projection. And here we have an important assumption: you are involved in an OOBE (at least to a degree) whenever you dream. What sets it apart from a full OOBE is your hazy consciousness during the experience and poor recall afterwards.

121

Teleportation: From Star Trek To Tesla

Many people forget most of their dreams completely. Learning astral projection requires a kind of inner mental clarity and alertness. Dreams are a door to the subconscious which can be used for psychological and spiritual insight, and sometimes for precognition. Dream content is influenced by external sounds and sensations. For example, a loud external noise (such as an alarm clock) will likely appear in your dream.

Dreams are also influenced by events of the previous day, by your moods, and by suggestion. Everyone normally dreams 4 or 5 times a night. The longest dreams occur in the morning. Everyone dreams. You are more likely to remember the details of your dream when you first wake up. By keeping a dream diary you will improve dream recall. Have writing equipment or a tape recorder at your bedside for this purpose.

Suggest to yourself several times before you go to sleep: *I will awaken with the knowledge of a dream*. Then when you do awaken, move quietly. Remember first, then write the dream down, and then add as many details as possible. The next day check for objective facts and expand if you can. Once you start remembering your dreams in this way, it will become easier to do so.

Different Forms of Teleportation

Teleportation may be subdivided into four basic types: **Physical Projection – Mental Projection – Astral Projection – Etheric Projection**.

Mental projection is really simple clairvoyance (remote viewing), and traveling in your mind. Imagination plays a key role. The experience of mental projection is not particularly vivid, and you will more likely be an observer than a participant. Nevertheless, mental projection is an important way in to astral projection proper.

During mental projection and astral projection you are able to travel through solid objects, but are not able to act directly upon them or to teleport them (if they are in the physical world). This is not true during etheric projection. Whether it is simply subconscious expectation, or whether it is a true etheric projection which in theory means that part of your physical body has been relocated with your projection (the etheric or vital part) may be difficult to determine.

Etheric and physical projections generally travel at or very near the physical world. There are even cases reported in which the entire physical body is transferred to another location, or cases in which the physical body exists and acts in two separate places at once.

Astral and mental projection are not confined to the physical world. Travel in the mental and astral realms is feasible, and often preferred. Nor are astral and mental projection restricted to the realm of the earth (you can even travel out into deep space to visit the planets and galaxies).

Teleportation: From Star Trek To Tesla

The electrical activity of the brain has been observed and classified with EEG (electroencephalograph) equipment; signals picked up from the scalp by electrodes, then filtered and amplified to drive a graph recorder. Brain activity has been found to produce specific ranges for certain basic states of consciousness, as indicated in 'hz' (hertz, or cycles/vibrations per second): delta – 0.2 to 3.5 hz (deep sleep, trance state), theta – 3.5 to 7.5 hz (day dreaming, memory), alpha – 7.5 to 13 hz (tranquility, heightened awareness, meditation), beta – 13 to 28 hz (tension, normal consciousness).

As you can see, some form of physical relaxation is implied in the alpha, theta, and delta consciousness. These states are in fact reached through deep breathing, hypnosis, and other relaxation techniques. OOBE occurs during these states, and delta is probably the most important.

The problem is maintaining mental awareness and alertness while experiencing these altered states. Experimental subjects hooked to an EEG do not show a discrete change from drowsy to sleep; it is very gradual. At the threshold between sleep and waking, consciousness is a drowsy condition known as the hypnogogic state. OOBE seems to occur during this state, or a variant of it. By careful control of the hypnogogic state (not going beyond it) it is possible to enter OOBE directly.

Basic Techniques

Most methods of astral projection are methods of conditioning. Some form of trance or altered consciousness is always involved. Although there are many techniques used to produce an astral projection, they boil down to nine tried and true methods. These methods, in helping induce a proper frame of mind for OOBEs, are also essential for learning to teleport. Care must be given not to rush through any of these techniques, as practice and patience are important for a successful OOBE or teleportation.

Diet – Though it is open to debate, there are those who believe that certain dietary practices may aid in OOBE and teleportation. These include fasting, vegetarianism, and in general the eating of light foods. Carrots and raw eggs are thought to be especially beneficial, but all nuts are to be avoided. Over-eating should be avoided. And no food should be eaten just before an OOBE attempt. If you intend to practice during sleep, for example, allow two to four hours of no food or drink (except water) before bedtime. In general, we see here the same kind of dietary restrictions advocated for kundalini yoga.

Progressive muscular relaxation – This is one of the basic methods used in hypnosis and self-hypnosis. Physical relaxation can assist one in attaining the requisite trance state. These techniques involve beginning at the toes and tensing, then relaxing the muscles, progressively up the entire body.

Teleportation: From Star Trek To Tesla

Yoga and breath – Yoga, mantra, and breathing exercises similarly aim at physical relaxation. The practice of kundalini yoga is particularly relevant, since it is concerned with altered consciousness. In fact the arousal of kundalini requires a similar state of consciousness to OOBE and/or teleportation.

Visualization – This involves a type of extended clairvoyance or picturing of remote surroundings. If you can experience the feeling of being there, so much the better. Although this technique is essentially mental projection, it is possible to deepen mental projection into astral projection through visualization. Aleister Crowley taught a similar technique: **(A)** visualize a closed door on a blank wall, **(B)** imagine a meditation symbol on the door, **(C)** visualize the door opening and yourself entering through it. J.H. Brennan describes similar techniques wherein the door is visualized, shaped and colored like a tarot card, and the student visualizes entering into it.

Guided imagery – In many respects similar to visualization. Except in this case, there is a guide (or perhaps a voice on tape) directing you by means of descriptions. As with visualization, mental rather than astral projection is most likely.

Body of Light – The old Golden Dawn technique. Imagine a duplicate (mirror image) of yourself in front of you. Then transfer your consciousness and sensation to the duplicate (body of light).

Strong willing – This is somewhat like creative visualization experienced in the present. You express your strong desire to project through sheer willpower while you visualize yourself doing it.

Audio Tapes - There are several good audio tapes on the market such as Brad Steiger's *Mastering The Art of Astral Projection* (#14) and Diane Tessman's *Astral Projection Techniques by Tibus*. Both distributed by the publisher of this book – Inner Light/Global Communications.

Dream control – This is one of the most important techniques. It involves becoming aware that you are dreaming. There are several ways to do this. Oliver Fox said to look for discrepancies in the dream to realize you are dreaming. One occult student visualized a white horse which he could ride wherever he wished to go. After a time, when the horse appeared in his dreams it was his cue that he was actually dreaming/projecting.

Don Juan tells Castaneda to look at his hands while he is dreaming. Even the tarot and Cabala may also be used as dream signals. Another method is to tell yourself each night as you go to sleep, "I can fly," then when you do, you will know you are dreaming. Once you know you are dreaming you can control your dream/OOBE and go anywhere you want.

Repetitive activities will also likely influence your dreams. For example, if you are on an automobile trip and spend most of the day driving, you will probably dream about driving. You can condition yourself to be aware you are dreaming by doing a repetitive activity many

Teleportation: From Star Trek To Tesla

times (walking across the room or a particular magick ritual, for example). Then when you dream about it, you will know you are dreaming.

Although all these techniques may appear straightforward, they all require effort and perseverance. Astral projection and teleportation are generally learned, though there are some people who appear to be "naturals" and can achieve their goals almost right away.

To some, astral plane is the "spirit world" into which one passes after death. It is possible for mediums to visit with the dead, or you might be called upon to reassure and help those who have just passed over (died) or those who are consciously projecting for the first time.

The astral plane could be the "in-between" place that teleporting objects and people go through to circumnavigate our 3-D world. Those who have visited the astral plane claim that time and space have no meaning there, and that they can travel anywhere and at anytime to our world through the world of spirit. In science-fiction this space is called "hyper-space," which is a space that is outside our own spacetime, and allows for almost instantaneous transportation from one point to another in normal space.

Dr. Bruce Goldberg, author of the book *Time Travelers From Our Future* (Llewellyn Publications 1998), writes that a true teleportation occurs when the physical body dematerializes (disappears) from one location and subsequently rematerializes (reappears) in a different spot in an instant, often accompanied by a "pop" sound as the surrounding air is displaced.

Goldberg believes that teleportation can also be used for time travel. In *Time Travelers From Our Future*, he states that the early time travelers (from the 31st to the end of the 34th centuries) required some form of craft to travel back in time. It is during the 35th century that teleportation techniques will be developed and ships are no longer required or utilized.

As was stated earlier, we can experience teleportation during the dream state. It differs from regular dreams and lucid dreams in that your body physically leaves the bed and travels to another location on a different dimension.

If you are observing someone actually being teleported, you would see their body slowly fade away and disappear. Nothing else in the environment would be altered. The person undergoing teleportation would experience an increase in their energy vibrating at high speed, accompanied by a tingling, buzzing sensation and/or feeling of spiraling upward.

Teleportation can take place in three different ways. The first is a spontaneous teleportation. You may have had the experience of performing some chore that you had done thousands of times before and sensing for a moment that you were someplace else. Another possibility is walking down a street, or driving a car and realizing that you are much farther from your original location than was possible over the period of time in question.

The second type of teleportation is one that is directed during your sleep state. This experience differs from regular lucid dreams in that you are not confronted with symbols or merely your subconscious transported to the dream world. Your physical body during teleportation has physically relocated to the astral plane or another dimension. This dimension is outside what we call normal spacetime, so travel through it is instantaneous. The memory of a true teleportation is clear and accompanied by recall of physical sensations. Most dreams are hazy and leave us confused as to their meaning.

A consciously directed teleportation comprises the third group. In the beginning you will find that you wind up in places far from your goal. This low level of accuracy improves dramatically with regular practice. When you return to your place of origin, you will again hear that "pop" sound. Your body travels at the speed of light, so distance is of no significance.

In order to successfully experience teleportation, you need to focus your mind and block out all other distractions. It is only your limiting beliefs that will prevent you from experiencing this unique form of travel.

Your body must be well rested to assure success. Because of this you will find it easiest to teleport at night. Since your body has had many spontaneous teleportation experiences while dreaming, the night is a natural environment and time for the novice teleporter to free themselves from the bonds of reality.

Goldberg recommends that you teleport to secluded places to avoid shocking other people when you materialize in front of them. Always act as if you are already in your desired location when practicing teleportation. See yourself as if you were viewing this new place from inside your body (first person perspective), not as if it were a dream or a movie. Make sure you see, feel and sense yourself in this new spot.

The natural tendency is to hold your breath during a teleportation. You must resist that temptation and breathe consciously by taking slow, deep breaths. However, don't hold it in, breathe normally and don't hyperventilate.

Since teleportation is a mental activity, we must not deprive our brain cells of its needed oxygen. You will feel lightheaded, with a swirling and upward spiraling sensation during this process. Occasionally, people report nausea during their initial stages of relocating. This will disappear quickly and be prevented by proper breathing.

Others report a sexual sensation in their body. Sometimes a feeling of tightness across the eyebrows (third eye area) is described. Your perception at your desired location will appear as if you were looking at your location through a keyhole. This disappears with practice. Always state clearly your intention to travel to a certain location or visit a certain individual. You do not want to imply that this is a desire that can fail, but a declaration of an empowered soul who is confident about succeeding in this venture.

For example, state to yourself, or out loud if it feels necessary: "I intend to teleport to Paris." You must always trust yourself and you Higher Self and acknowledge your ability to teleport anywhere in the universe or to any other dimension.

Teleportation: From Star Trek To Tesla

Dr. Bruce Goldberg's Teleportation Exercise

Try this exercise in teleporting to visit a friend. Visualize the face of the person to visit. See only the face and try to see it in clearest detail. As you hold this image in your mind's eye, you will at first begin to remember the clothes the person wore, how they acted, what they said and did, but pay no attention to these recollections.

Simply focus in on the face, not the body or actions and gradually these impressions will begin to fade. As they do, you may pick up a tiny spot of light, or maybe a large one, somewhere away from the face but definitely in your line of vision. Try to focus in on it.

Usually it will seem fairly close to you, almost as if it were suspended in midair about ten to twelve inches from your eyes. When you perceive this "intruder" on your otherwise clear picture of the face of the person you are trying to reach, concentrate upon it.

You will find that it will appear to expand and include more detail, as locations like a room, or an automobile, or a stadium, and in this location you will see the person you are trying to reach. See them as they actually are and in their exact surroundings at that precise moment. Remember, you cannot bring this in by force of will. Once you see the other person's face clearly, relax a little. Try to be passive rather than active. You are trying to receive an impression, not create one. So let it come in. Stay here as long as you feel comfortable before returning to your practice room and ending this trance.

This exercise is presented in Goldberg's *Time Travelers Training Program* cassette album, which instructs you to explore the fifth dimension, meet time travelers from our future and teleport your physical body anywhere on this planet and to other dimensions safely and return to your present location. (www.drbrucegoldberg.com/index.html)

Another valuable source of material on teleportation is a five volume series entitled, *Life and Teaching of the Masters of the Far East*, by Baird T. Spalding (DeVoss & Co. 1935). Volumes three and four are especially helpful for mental preparation.

Spalding claims to have witnessed and experienced the art of teleportation by the masters of the Far East. One of the members of his party learned to teleport after observing how the masters seemed to appear and disappear at will. Feeling frustrated about recent experiences his team had been through, Spalding listened to a member describe his disheartening feeling after observing a human act of terrorism against fellow humans.

"'Every one of us has taken upon himself the condition of the experience through which we have passed. This is what is now hampering us and I for one am through with that thing, it is no part of me whatsoever. It is not mind only as I worship it and hold to it and do not let it go. I step forth out of this condition into a higher and better condition and let go I am entirely through with it.' As we stood and stared at him, (Spalding reported), we realized he was gone, he had disappeared."

Teleportation: From Star Trek To Tesla

Feel The Energy

Here are some of the concepts I put into practical use while in the military's secret training program.

- ❑ As you continue to do your teleportation exercises, it will become easier to visualize yourself at your target location. You should soon begin to not only see yourself there, but you will also begin to experience other sensations: the smell of the air, the sound of the wind or birds singing in the trees, the feel of the ground beneath your feet. You will notice that your body feels as if it is spiraling upward. Your energy twirls around, moving upwards into the sky. Although others near you will not see it, you will feel your energy swirling within you, leading you to sway a bit, or to feel somewhat lightheaded. It is important to remember at this point to continue breathing in slow, deep breaths.

- ❑ When you first begin to experience teleportation, it is not uncommon to feel that you are seeing your location as if through a long, narrow tunnel. Your view will expand as you practice. Also, it may feel as if your entire body did not make the trip, that only segments of your body teleported.

- ❑ Another common experience is to feel as if you are there only in the blink of an eye. Again, the more you practice, the more complete your experience will become. You will soon see not only the entire location, but you will feel and smell it as well.

- ❑ When you first begin teleporting your sense of direction may not be the best. This happens to most teleporters in the beginning, so don't worry about getting lost. As you practice, your skills will sharpen and you will arrive in the desired place. Relax in the knowledge that when you are first learning, you will automatically return to your place of departure.

- ❑ Your teleportation experiences will magnify over time. You may have several practice sessions with a glimpse or two of success, and then in the next session you may have a major breakthrough or nothing at all.

- ❑ It is normally a good idea to start small when attempting to teleport. Don't try to reach Mars for your first trip. Instead, aim for someplace familiar and nearby, say your backyard or another room in your house.

Teleportation: From Star Trek To Tesla

As you develop further in teleportation, you will find that it is easier, and almost automatic, to go to different places. When you have learned to control your abilities, you will find that distance has no bearing on where you want to go, next door or across the galaxy is the same distance in the world of instantaneous teleportation. Your mind will be your guide, and you will find that a fleeting thought can send you careening unexpectedly to an unintended location. So learn to control your thoughts.

Soon, your neighborhood, your town, your state, the entire planet could be within easy reach to your teleportation abilities. Start out small, but then feel free to explore the places you have always wanted to visit.

The universe is now yours to explore. Go to it.

CHAPTER 13

PERSONAL EXPERIENCES WITH TELEPORTATION

Whether spontaneous or intentional, personal experiences with teleportation are not all that uncommon. Most people merely shrug off their occurrence and forget about it, rather than risking ridicule by telling their friends and family. Few people realize that others have had strikingly similar experiences.

The Malaysian Star newspaper reported on February 20, 2001, in the town of Kampung Kepayan Baru, a woman told police that her husband became "invisible" right before her eyes and disappeared.

Keningau district police chief Deputy Supt Abdul Hadi Baharudin confirmed that a report was lodged by a housewife who said her husband, hospital assistant Yabi Gintukod, age 45, had gone missing. However, he declined to give details of the police report, only saying that they have classified it as a missing persons case.

"Like all missing persons cases, we have flashed the information to all OCPDs and are seeking the help of the village committee to help trace Yabi," he said.

Mainis Gampat, a mother of eight, claimed that Yabi suddenly became invisible and vanished into thin air before her eyes just after dinner on Feb 20. "We have not been able to locate him since then," she told reporters at her house in Kampung Kepayan Baru.

She claimed it was the second time that her husband had disappeared. In the last incident which occurred two days before his second disappearance, Mainis said Yabi was found sitting in a bush in a stupor. They sent him for a medical check-up and he was found to be in good health.

Yabi's brother-in-law Mahat Kulimpang claimed that Yabi had spoken to him of an encounter with a man whom he claimed to be an "alien" with a square body. "Yabi told me that the alien wanted him to go to a strange place, and that it could teleport him away anytime it wanted," Mahat said.

He claimed that Yabi had always wanted to go to that place. "The night before Yabi disappeared, he was wearing white and his feet were not touching the ground. "When I touched his shirt, it slowly turned black and he soon passed out," claimed Mahat, who seemed puzzled by the phenomenon.

Fourteen days later, Yabi stumbled out of a forest about twenty miles from his home. He was disoriented and did not realize that two weeks had passed. Yabi could offer no explanation on where he had been, or how he had come to find himself so far away from his home. "I remember being at home with my wife, when I suddenly felt myself disappearing and spinning into the air, and that's when I found myself lost in the woods."

Teleportation: From Star Trek To Tesla

Things That Vanish and Reappear

Paul Whitehorn of London, England reported that in 1981 he and his girlfriend were asked to "babysit" two Siamese cats at the house of a friend. They were instructed to put the cats into the walled in backyard if they decided to leave the house.

About an hour later my girlfriend said she wanted to go out and I went to collect the cats to put them in the back yard. One of them, a very strange and vocal little animal, refused to be gathered up and ran to the foot of the stairs. As I approached the cat it darted up six or seven steps and waited, calling out loudly and looking at me but, as I got near again, it ran up some more steps where it stopped again and sat staring at me, mewing loudly.

To my astonishment and annoyance the animal did this repeatedly as though it were teasing me until we had ascended two flights of stairs and reached the third floor landing at the top of the house. "Now I've got you" I thought as I reached the landing because I saw the cat dart in through the slightly open door of a small disused room. Approaching cautiously, bent low with hands outstretched to stop the animal running out of the door and past me, I entered the room, carefully closing the door behind me.

The cat was nowhere to be seen! The only contents of the room, which measured about nine feet by eight feet maximum, were an old armchair, an empty crate, a cupboard with a locked door, and a few odds and ends. Thinking the animal had hidden itself I began to search through the few available hiding places, even unlocking the cupboard and looking inside. I turned the armchair upside down to see if the fabric was torn allowing the cat to find its way inside but there was no sign of any place where an animal could hide. I looked up the chimney (it was blocked by an iron plate), looked into the crate and searched through the room thoroughly for about half an hour. The one window was locked and there were no other doors or exits available.

Eventually my girlfriend, exasperated at the amount of time I was taking, called up the stairs to ask what I was doing. I had searched the room thoroughly three times and eventually gave up, feeling baffled and irritated that I was unable to carry out my friend's request. As I left the room I closed the door behind me determined to search the room again on our return.

We went out but later, on our return, I told my friend's wife what had happened and apologized to her for failing to put the cat out. She stared at me and said, "That's alright. When we returned both cats were out in the backyard calling to be let in."

I have told this story of the dematerializing cat several times inviting some explanation but none has ever been forthcoming the usual response being disbelief. To this day I have no explanation for how the cat got out of the room without me seeing it.

Teleportation: From Star Trek To Tesla

Sammy Mitchell of Lynchburg Virginia, relates an interesting case of a teleporting watch.

I am a 48 year old woman who has lost both elderly parents within the last year and a half. My father passed away first and shortly after his death, I visited my mother who had many serious medical problems.

On this visit, I was getting ready to bathe her in the bathtub. I was the only one she trusted to do this. Anyway, I took off my rings and watch and put them down somewhere in the house; either upstairs or down. Hours later, I went back to the bathroom and picked up my rings and returned them to my hand. However, I couldn't find my watch.

I searched the whole house. First downstairs, then upstairs. I went to the upstairs bedroom and walked to a bed. I lifted clothing I had left there in search of my watch. I remember saying to myself, "I wonder what I did with it (the watch)?"

I took the ten or twelve steps to the bathroom and let my eyes walk around the room. No watch. As I turned around, I saw my watch lying on it's side on the very bed I had searched so carefully only moments ago. I feel this is a way that my dad chose to let me know he was still around. I often smell his presence in the house from time to time.

A woman who identified herself only as Natasha from the United States was afraid to tell her story of hauntings and teleportation because she didn't want to be thought of as crazy.

Some people will believe this, some won't. I can assure everyone it is true. I lived in a house in Pennsylvania where strange and scary things happened all the time. For example, I was getting ready to go out one day, and I plugged in my curling iron in the bathroom on the second floor. The only other person in the house was my mother; she was downstairs. I went down to talk to her for a few minutes and went back upstairs to curl my hair. The curling iron was gone. IT WAS GONE! My mother and I searched everywhere.

I was so mad I kept looking for about two hours and so did my mother (this was not the first time something had just vanished). I tore apart drawers and closets, looking everywhere. We finally found it downstairs in the dining room closet underneath a huge pile of coats and other junk. The cord was wrapped neatly around it.

One more example. I was at home alone one day. Just me. I put my drivers license down on the kitchen table and when I went back to get it, It was gone. Again I searched the house up and down. I never found it. Finally, after about a month, I went to get a new one. I came home with my new license, and sat down on the couch (and I was alone again). I watched TV for a few minutes and then I looked down at the cushion beside me. There sat my license.

I have quite a few stories like that, and some are on the scary side, but I haven't told hardly anyone. Who would believe me anyway? They'd just think I was a nutcase.

Teleportation: From Star Trek To Tesla

In his book *UFOs Among The Stars* (Global Communications, 1992), Timothy Green Beckley says that punk rocker Helen Wheels "gives a totally 'no nonsense' performance; her antics as part of the New York punk rock scene is most legendary. I understand she used to open beer bottles with her teeth."

Helen Wheels (who unfortunately passed away in 2000), was probably best known in the music industry for songs such as *Sinful Love*, and *Tatoo Vampires* that she composed for the popular seventies group **Blue Oyster Cult**. Helen proudly displayed several gold records for her writing accomplishments in the entranceway to her Manhattan apartment.

Her closest ally was her pet snake Lilith, a 95-pound Colombian boa constrictor who used to wrap itself around Helen's neck as the band blasted out some fairly heavy numbers behind her. Lilith the snake was the culprit behind this interesting teleportation story as recounted by Helen's brother, Peter Robbins.

About 10 or 12 years ago, Helen and I were sharing an apartment on West 57th Street. Her boa constrictor, Lilith, lived with us, along with a smaller snake and our cats, Wolfie and Dweezel. Lilith was a large boa and measured 9 or 10 feet at this time. She lived in a 4' by 4' by 8' hutch that had sliding doors on the front. The Hutch was made of wood with holes strategically drilled for air to circulate. The doors, which were built into a pair of groves in the wooden structure, were cut from quarter inch clear plastic.

On this particular afternoon, my friend Jim was visiting from Oregon, and in fact was napping on the couch which stood opposite the snake's hutch. I was working in my room when I heard Helen call my name to come into the living room. I walked in to find Jim just waking up, and our large tiger cat Wolfie sitting about ten feet from the hutch, literally still as a statue, all puffed up, and 'pointing' at Lilith, who now lay on the carpeting in front of her hutch.

As we entered, Wolfie began to back away very slowly and Helen, Jim and I began to speculate on how Lilith could have gotten free. The snake, who was very good with people, was easily coaxed back into her home by Helen and me.

Yes, the doors had not been touched and both were still in their closed positions. And I guess it is tempting to go for a paranormal explanation. However, snakes are particularly clever in terms of escaping their containers as we had experienced and observed over the years with other such critters. They are also one large muscle which has the ability to push into the smallest spots with their noses and then work the rest of themselves through an increasingly expanding opening – even through a space that began as a quarter inch or so separating two large pieces of plexiglass.

It would have taken some time and effort, but Lilith had plenty of both. Teleportation? Maybe. But, hey, none of us were watching, so this one will have to remain a mystery.

Helen Wheels with her pet snake Lilith

© Mariah Aguiar - www.specialforcesmedia.net

CHAPTER 14
THE TELEPORTATION OF SACRED STONES
BY DIANE TESSMAN

Diane Tessman was raised in Iowa where she had an abduction experience – with strange markings to prove her experience was legitimate. Years later she started channeling her main cosmic confident, Tibus, who she describes as a time traveler. Her books include: *Earth Changes Bible*, *7 Rays of the Healing Millennium*, and *Passport to Heaven*, co-written with Tim Beckley.

While I lived in Ireland, I visited as many ancient standing stones and dolmens as I possibly could because these ancient, majestic, and mysterious megalithic objects take my breath away each time I encounter them.

They stand alone in the rainy, cool Irish countryside, often marked only by a small signpost. They are usually in a remote field with no one around for miles; one parks the car and walks a mile or two to reach them. And suddenly, there they are, standing alone, strong, and infinitely mysterious, just as they have stood for six thousand years. What history they have witnessed! Four thousand years after they were created, Jesus walked on Earth.

There are well over 1200 sacred stone sites in Wales, Scotland, Britain, and Ireland. What precious links to our ancient spiritual past, and perhaps to our spiritual future.

They still stand, despite the fact that many of them were ransacked and purposely destroyed by a variety of conquerors and Christian zealots. The ancients must have erected many thousands of these spiritual sites originally.

In a delightful variety of ways, movements of the sun and moon, solstices, and other astronomical events are charted by the positioning of specific monoliths and groups of stones in the British Isles. "Adam and Eve" are two stones at Amesbury, Wiltshire, England, which remarkably symbolize the miracle of creation through male and female genitalia configurations, but even these two stones are aligned to also give astronomical information.

At Newgrange in Co. Meath, Republic of Ireland, there is a "window box" above the entrance to the inner chamber into which the winter solstice sun has beamed through, at the crack of dawn on December 21st, for at least six (perhaps seven) thousand years; Newgrange is thought to be over a thousand years older than the Great Pyramid in Egypt.

The solstice sunbeam enters Newgrange's window box and travels down the long tunnel-like entrance, striking the back wall of the inner chamber. The ancients positioned it to do precisely this, for eternity.

Modern science at first thought Newgrange was simply a tomb for royalty; often they have

**Above: Diane Tessman
Below: The Harristown Dolmen**

ignorantly said this about mysterious ancient sites. In fact, it has since been proven that Newgrange was much more: an astronomical observatory, a spiritual oasis to mark the darkest day of the year and to crystalize the hope for Spring. If one holds a crystal to this solstice sunbeam, the entire chamber lights up as if by magic.

Newgrange may have other functions and significance which we do not yet comprehend. It's construction baffles modern engineers who try to figure out how it was built. For these many thousand years, Newgrange has never leaked a drop of water, yet it was built of free-standing stones which rest one upon the other in an ingenious pattern. It stands perfectly dry in the extremely rainy and always damp Irish countryside near the Boyne River.

I am convinced that the ancients knew the secret to teleportation when these chambers, standing stones, and dolmens were erected. I have stood and marveled at the least known, most remote dolmen in the Irish countryside as well as the more famous sites such as Stonehenge, and reached the inescapable conclusion that the gigantic top stones which balance so precariously on another huge stone at one tiny intersect point (and have done so for six or seven thousand years), were put into place through teleportation, not manual or horse power, nor any other physical means.

There are a variety of metaphysical theories as to how this was done, such as placing the cap stone weighing many tons, on top of the other stone through secret, mystical musical chords which produce levitation.

It is said that the shaman of the ancient pre-Celt people, knew the right combination of notes and chords. To the shaman, this may have been a secret which was already old in her day, handed down from the obscure civilization of Lemuria or Mu.

Or perhaps this secret was given to the shaman by the Sky People who are said to have been the first tribe to inhabit the island of Ireland. These were the Tuatha de Danann (Family of Diane). Or did the Sky People themselves erect these ancient stone sites?

Perhaps it was not secret musical chord combinations but advanced technology which teleported these gigantic stones into such unbelievable positions. If this is so, was it unknown UFO occupants whose spacecraft had tractor beams for levitation?

Or was it the Sky People's starships which had teleporters that rearranged the molecules of stones and teleported them into place? Perhaps the Sky People still possessed their starships after they crashed or chose to settle – on Earth; then over time, the ships and their technology were lost, but by then their sacred stone sites were completed.

Another theory: the ancient stone sites may be monuments to a pre-time when technology and building materials were highly advanced. Are they memories in stone which echo back to an even earlier time?

Perhaps after global cataclysm, there was only stone left to use as a building material. The concepts of the sites, such as the astronomical observatory complexities, seem to be more advanced than the material (stone) which was used. Yet, the creators persevered some how, and so their stone holy places and stone astronomical observatories stand strong to this day.

One fact is certain: The sacred stone sites possess secrets and power which our modern minds cannot yet grasp. But our spirits can and do perceive the mystical impact and intensity of these holy places.

Our spirits are as old and as new as these magnificent stones, and just as magical. We can teleport ourselves to them through creative visualization, and we can revel in their awesome power as we explore their mystery. Maybe they can even help us survive into our future.

Channeling from Tibus: Two Kinds of Teleportation

Diane Tessman has channeled her star guardian Tibus for nineteen years; together, Diane and Tibus have counseled thousands of clients with steady, accurate predictions and guidance regarding Earth changes and disasters, offering a road map to surviving into a new dawn. Tibus describes himself as "a voice from Earth's past and Earth's future."

This is Tibus. I come to you in love and light.

When Diane started exploring her spiritual roots and identity, and her psychic gifts of channeling and precognition, she looked to the stars for the answer to my identity. You see, I was this voice – a presence – whom she felt; I was not (am not) a literal voice within her head.

As her explorations progressed, she remembered my presence during two childhood UFO encounters. She had been outside on a cold evening in November on her parents' farm in north Iowa, and suddenly she was aboard a star craft. She did not realize it was a star craft at the time, but she was instantly "inside" in a place she did not know. She felt no fear. "How did I get on board?" she asked me years later. "Our craft has the ability to teleport," I answer, "kind of like good old Star Trek."

In time, Diane began to be psychically pulled to ancient Earth times and places. Her original fascination with the stars, space travel, UFOs, and the future, remained part of her, but she realized that there was more to the picture: More to learn, more to know about me (Tibus) and ultimately, more about herself. Like all seekers of truth, she began to look to Mother Earth and to her mystical past.

She was inspired to move to Ireland in 1990 where she explored the dolmens, standing stones, and ancient structures such as Newgrange. The land of Ireland (Mother Ireland) spoke to Diane, as she speaks to all who will listen, and Diane learned more about our sacred planet Earth, about the precious life-forms (including humankind), who are her children, and her creations.

In Ireland, Diane explored a spirit presence she named "Fallon." She felt his anger, his love of Ireland. In time, I told her my story. Yes, I was Fallon, I am Fallon. In a nutshell, I (Tibus) was Fallon in a past life. It is a bit more complicated then that, but matters of space and time are impossible sometimes to get into meaningful words.

At any rate, I told her Fallon's story and in time, we managed to help Fallon out of the vicious circle he had fallen into; It was like freeing a ghost who had been trapped by its own fears so he could progress to a new life.

Diane asked me, "Tibus, how did you get back into time when Fallon lived his life?"

"All souls are travelers of space and time, Diane. Believe it or not, we all teleport around the universe in an awesome spiral, landing at different points, experiencing this lifetime or that lifetime."

"Is it the same teleporter you used to beam me aboard that cold November night?" She asked.

"No, that teleporter is a mechanism of advanced science. It beams your molecules from one point to another. But souls teleport around the universe solely on the power of their spirit; this is the reality of consciousness itself. There is no mechanical or scientific device involved. We all have the power to do this and, we all do it."

Tibus continued: "We all have an amazing variety of aspects (lifetimes) which add up to our Whole Self. When a soul reaches the point of true enlightenment, he or she becomes the Whole Self. Then he can look back on every lifetime he has experienced, and remember it with full knowledge of all other lifetimes, also cognizant of he who is now whole.

"I am simply a being who has attained this wholeness, and so I have the ability to come to you as Tibus or Fallon. It is my responsibility, as it is with others like me, to give you guidance and knowledge as you progress along your path. It is like an older brother holding out his hand to help you up the steep cliff. This is what angels do, this is what guides do, this is what star guardians do, this is what higher beings do."

Tibus is always anxious to connect the spiritual and the scientific, and so his final thoughts on this subject: "If you think I am babbling metaphysical concepts, look to the field of quantum physics. It will tell you that there are indeed particles which 'bypass' the speed of light; they are literally in two places at once. In fact, they are in more than two places at once. Consciousness particles teleport themselves instantaneously; consciousness knows no such thing as 'time,' it is not a prisoner to a point in time. There are no physical barriers to consciousness, it is not restricted to one body. All are one.

"May the light of goodness surround you, always, **Tibus.**"

Spoken Through Diane Tessman, P.O. Box 352, St. Ansgar, IA, 50472
Ask for a free sample of her newsletter and visit her website at: www.dianetessman.com

Teleportation: From Star Trek To Tesla

FROM THE SOUL OF NIKOLA TESLA:
Bi-Location and Teleportation Methods

An Audio Cassette Companion To The Book You Are Now Reading!

Back while she was growing up in Iowa, Diane Tessman had an experience she will never forget. Playing in her yard she felt compelled to walk off by herself to an isolated area. It was here, out of the way of the eyes of family members, that Diane felt herself being teleported to another "place and time," which she relates in her first book *The Transformation*. For lack of a better term, Diane best describes her experience as being UFO related – but she has always maintained that the beings she met might not have been from outer space, but were probably time travelers.

Since her original experience, she has remained in contact with one of the beings onboard the craft who is very human in appearance: "Probably because he is from Earth's future," the blonde haired author of *Earth Changes Bible* explains. Tibus, as she identifies him, has made numerous remarks about Earth's destiny and the safety of the planet, "no doubt because what happens in our time and in the next few years will ultimately have a definite impact on the future survival of the human race."

Recently, Diane moved from Ireland back to Iowa where she purchased the land she grew up on. In the last few years she has had several other experiences on the property. One remains foremost in her consciousness. "I was drawn one evening back to the spot where I had first been teleported on board a craft which we can call a UFO for lack of a better term. There was this mist or fog in the field, and there was a man standing in the fog. He was dressed fairly conservatively and looked very dapper."

Toward the end of the conversation the man revealed his identity. "He said he was Nikola Tesla and that his work was not finished on the earth plane and that he had to get a message across to those who were willing to listen."

Diane reveals in a specially prepared audio tape, made exclusively for readers of this book, how Tesla told her "that at the time of his death he was working on a revolutionary discovery that included his personal – previously unpublished – secret 'mental process' that will enable any individual to engineer their own reality and help develop what he calls 'bi-location,' a mental form of teleportation.

"Tesla," Diane further explains, "sees this as his ultimate gift to humankind and to the planet which he loved so much and felt sorry about having to leave."

The 60 minute cassette is a training tape for those who wish to explore the many possibilities that are now open to us all in the New Millennium thanks to the great genius that was – and still is – Nikola Tesla! Readers wishing to obtain this tape may do so by sending $10 plus $2 shipping directly to the publisher, Global Communications, Box 753, New Brunswick, NJ 08903

Ask for Diane Tessman's – *FROM THE SOUL OF NIKOLA TESLA: BI-LOCATION AND TELEPORTATION METHODS.*

Order all your Tesla products from:
www.TeslaSecretLab.com
24 hour automated order hot line (732) 602-3407

TESLA'S "LOST" INVENTIONS AND THE SECRET
OF THE PURPLE HARMONY GENERATORS

Upon the passing of Nikola Tesla, huge boxes containing his private journals and unpatented inventions were retained by the Custodian of Alien Properties and were locked away. From inside information gathered in the years following his death, it was ascertained that officials from Wright-Patterson Air Force Base (also the home for many years of Project Blue Book, headquarters of the government's UFO cover-up attempt) hurried to the warehouses of the Custodian of Alien Properties and took possession of all of Tesla's documents and other materials, all of which were classified at the highest level. To this day, a great deal of Tesla's papers remain in government hands and are still highly classified. There are literally tons of notes, documents, drawings, and plans, as well as over twenty boxes of reportedly "missing" Tesla material. The government distributed false rumors that Tesla never kept notes, which is a blatant lie.

Over the course of time -- largely in the last decade -- some of Tesla's lost journals have been uncovered, and a number of his "secret inventions" have been privately developed. One of these inventions is Tesla's Purple Harmony Generator -- also know as Tesla's Purple Energy Plate. Though the "generators" have been around for a number of years, they are only now starting to receive the international attention they deserve in the alternative energy field. In an article published in the August 2000 edition of the popular FATE magazine, author Corrie DeWinter mentions that she first became aware of the generators while reading a book called STAR SIGNS by Linda Goodman. "Goodman mentions that the person who created the plates with Tesla preferred to remain anonymous. However, after the inventor's death, the company which produced the plates decided to give him due credit. The inventor, Ralph Bergstresser was born in 1912 in Pueblo, Colorado, of German parents who immigrated to the United States. He was extremely interested in free energy, or Zero Point Energy" as it is now called in scientific circles. Bergstresser carefully studied anything written about Nikola Tesla's experiments, and attended many lectures given by Tesla. At one point they were introduced and quickly became friends, due to their shared interest in free energy."

According to the FATE article, Bergstresser continued with his work for many years and following Tesla's death came into possession of several notebooks which helped him further develop the harmony plates. For all intents and purposes the plates look innocent enough. Coming in a variety of sizes, they are purple in color and are said to be "in resonance, or in tune, with the basic energy of the universe. They function as transceivers...creating a field of energy around themselves that will penetrate any material substance by osmosis. The energy is very beneficial to all life...plant, animal, or human. It might be considered as Positive Energy." Somehow or other -- according to current thinking regarding the plates -- the original atoms of the anodized aluminum structures are restructured when put through a proprietary process whereby the vibrational frequency of the atoms and electrons is changed.

Non-approved FDA testing has reportedly shown that the healing process is accelerated for burns and bone fractures when the injured party becomes the focal point of the purple plates' force field by wearing one of the self contained generators. Aches and pains are said to go away, the quality of sleep may be improved, water and food becomes more tasty (to establish this simply put a purple plate on a shelf in your frig. The quality of cheap wine is remarkably enhanced. Plates have been placed under sick houseplants, and near the food dish of small pets. Corrie De Winter in her FATE article offers several suggestions for the use of the plates: "Place a small-size plate in a pocket or purse for energy...small plate (is often) placed on forehead to alleviate headache pain, on joints to alleviate gout and arthritis pain, on stomach to stop nausea...Placed on forehead in the morning will help you to remember your dreams...I have also read testimonials from plate users who claim they help with cramps, headaches, stomach upsets, stiff joints, torticollis, swelling, ringworm, 'clicking' jaw, alcoholism, anxiety, colic and depression."

Probably one of the most influential tests has been conducted by the Perrysburg School District allowing them to stop using dangerous pesticides around the Frank Elementary School pupils and very naturally by utilizing the Tesla generators or plates. According to the school custodian, the plates where installed in the cafeteria and elsewhere around the building allowing them to greatly control the pest population. One of the most commonly used pesticides was developed by Hitler in World War II to penetrate mustard gas masks...the purple plates provide a totally safe means to attack the problem of pests. Tesla's Purple Harmony Plates come in a number of sizes and are now available through the publishers of this special insider report (see next page). Test them yourself!